An Introduction to
PHOTOBIOLOGY

AN INTRODUCTION TO
Photobiology

The Influence of Light on Life

by

YVES LE GRAND

Professor at the French Museum of Natural History, Paris
Vice-President of the International Commission on Illumination

translated from the French by

Michael Peckham, M.A., M.D., D.M.R.T.
and Lesley Williamson

FABER AND FABER
London

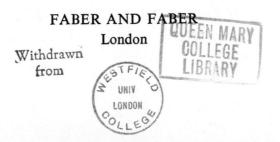

First published in 1970
by Faber and Faber Limited
24 Russell Square London WC1
Printed in Great Britain at
the Pitman Press, Bath

ISBN: 0 571 08849 X

Contents

CONTENTS

Preface

The English translation of this book follows the French text of *Lumière et Vie animale*, published in 1967, closely, apart from a few errors which were brought to my attention by my French colleagues, to whom I am very grateful, in particular to Dr. Raymond Latarjet in Chapter III and Professor Ivan Assenmacher in Chapter VIII. Some additions have also been made in order to keep abreast with advances made since 1967.

I am very grateful to the translators who achieved with success, as far as I can judge, the difficult job of clothing my thoughts in a British suit. It is a rather curious experience for an author to read a translation of his work and sometimes I found myself thinking, 'This fellow has funny ideas'. I then realized that I was reading my own text, but in a foreign idiom. This was a most agreeable experience.

YVES LE GRAND

Paris, January 1969

CHAPTER I

Light and Matter

1.1. Classification of radiation

Most of the information about the world which reaches the species *Homo sapiens*, to which the reader belongs, is visual. According to psychologists sight supplies 40 per cent of all sensory information to man. Next in order of importance is touch and finally, and a good deal less important, hearing and smell. It is of course a grave social disadvantage to be deaf, but the blind man is incomparably more cut off from his surroundings, and deprived of defence against them. In the theatre there has always been a distinction between defects such as stammering and deafness, which can to some extent be laughed at, and blindness, which is tragic.

The importance of sight and touch is implicit in the original description of the world: objects were seen by means of *light* and the *matter* of which they are made was handled or struck.

The sun was and is the chief source of light. Primitive peoples worshipped the sun as a divinity essential to both animal and plant life. Newton demonstrated in 1666 that sunlight is not a single entity. He directed a thin beam of light which shone through a hole in the shutter of his room through a glass prism, and established that 'white' light could be broken up into a continuum of coloured lights, with red at the end of the *spectrum* least deflected by the prism, and violet at the other. Orange, yellow, green and blue are ranged between red and violet (Newton added indigo, so that there should be seven colours as there are seven notes in the scale). These monochromatic lights are single, they cannot be split up.

1

Huygens' wave theory was enunciated in 1678 and experimentally confirmed by Young and Fresnel at the beginning of the nineteenth century. This made it possible to characterize monochromatic light by a measurable parameter, its wavelength.

Interference phenomena prove that light is a periodic phenomenon, and the wavelength its period in space, that is, the smallest distance between two points in the same phase along a ray of light.

Roughly, the wavelengths of the colours at the ends of the spectrum are 0·4 and 0·7 microns. ($1\mu = 10^{-3}$ mm $= 10^{-6}$ m). Table I gives the difference in *nanometres* (a nanometre (nm) is equal to 10^{-9} m; its old name was millimicron) between the main colours of the spectrum: these are average figures and obviously subjective, for normal observers.

TABLE I

Average limits between the colours of the spectrum

Ultraviolet-violet	$\lambda = 400$ nm	$w = 3\cdot10$ eV
Violet-blue	450	2·75
Blue-green	500	2·48
Green-yellow	570	2·17
Yellow-orange	590	2·10
Orange-red	610	2·03
Red-infrared	700	1·77

In 1800, Herschel found that a part of the sun's spectrum the infrared, extends beyond the visual red. Infrared is invisible and was discovered through the heating of a thermometer. The following year Ritter established that the photochemical action of blue and violet light on silver chloride is continued in the ultraviolet for wavelengths less than 0·4 μ. Visible light thus only forms part of the radiation emitted by the sun.

We shall not dwell here on the successive theories put forward by physicists to account for the properties of light, and the associated invisible radiations in the sun's spectrum. We shall only recall that light waves are, as Maxwell (1865) showed, electromagnetic waves in which the electric and magnetic fields at any point are perpendicular to each other and to the direction of propagation. Analogous waves, of much longer wavelength than light, were obtained by Hertz (1887), using purely electric techniques. These radioelectric waves can be produced over a considerable range of wavelengths, from less than a millimetre (thus linking up with infrared) up to several kilometres. X-rays, discovered by Röntgen (1895), by bombarding material in a vacuum with electrons, extend the range of short-wavelength rays beyond the ultraviolet, and are themselves followed by gamma rays, from radioactive bodies, and cosmic radiation. Light is only a very small part of the enormous range of electromagnetic radiation. It is those wavelengths to which the retina of the eye is sensitive.

All electromagnetic radiation is propagated in a vacuum at a speed c, of about 300,000 km/second; this speed is a very important universal constant. In matter, the speed becomes c/n, the refractive index n being usually greater than unity and varying with the wavelength λ (it is this *dispersion* which accounts for Newton's results on the splitting of white light).

Due to propagation a temporal periodicity is superimposed on the spatial periodicity measured by the wavelength. At a given point the *frequency* v, measured in *hertz* (Hz the reciprocal of the second), is the number of oscillations per second. The following relation exists between frequency and wavelength:

$$v = c/n\lambda \qquad (1)$$

Whereas monochromatic radiation is characterized by the frequency, the wavelength depends in addition on the refractive index of the medium under consideration. For example, radiation which strikes the retina of the eye must travel through a transparent medium (the *vitreous body*) whose refractive

3

index n is about 4/3; and therefore the wavelengths are $\frac{3}{4}$ of their length *in vacuo*, or in air ($n = 1\cdot0003$). It would therefore be preferable to define monochromatic radiation by its frequency rather than by its wavelength. λ has, however, been commonly used because its numerical values rest on a more usual scale: a fraction of a micron is a length which seems meaningful, because such precision is possible in industrial technique. On the other hand, the frequencies at the red ($0\cdot7\,\mu$) and violet ($0\cdot4\,\mu$) ends of the visible spectrum are $4\cdot3 \times 10^{14}$ and $7\cdot5 \times 10^{14}$ Hz, numbers which convey nothing to the imagination.

The *quantum* theory of radiation provides a much weightier argument in favour of frequency: introduced by Planck (1900) with regard to *black body* radiation (see below), and more precisely defined by Einstein some years later with regard to the photo-electric effect, this theory postulated that all energy exchange between matter and radiation takes place in a discontinuous manner, by multiples of a fundamental energy unit or *quantum*, whose value, for monochromatic radiation at a frequency of v, is written:

$$w = hv \tag{2}$$

Planck's constant h is a fundamental concept in physics. Though the propagation of light can be accurately explained by the wave theory, a *corpuscular* theory is needed to explain its emission or absorption by matter. The *photon*, the fundamental unit of radiant energy (w) reconciles Huygens' wave theory with Newton's emission theory.

For reasons which will appear later, it is usual to calculate the photon energy (w) in terms of the *electron volt* (eV). One electron volt is the kinetic energy acquired by an electron (the fundamental electric charge) subjected to a potential difference of 1 V. Taking this unit and calculating the wavelengths in nm for a vacuum (where $n = 1$) we have:

$$w = 1\cdot240/\lambda \tag{3}$$

What really characterizes a photon with respect to its behaviour in matter is its energy w, and as this is proportional to v, the conclusion that it is logical to distinguish photons by frequency rather than wavelengths is again reached. In the figures in this volume, photon energy (w) will appear on abscissae, but to facilitate comparison with the more usual notation, λ will also be shown on the scale, though it will obviously not be linear in λ.

1.2. Light Sources

Some artificial light sources emit what is known as a *line spectrum*, which means that the beam is made up of more or less monochromatic radiations, separated by intervals over which there is no emission. This is the case in mercury vapour lamps at low pressure, which under the action of an electric potential emit ultraviolet (254, 313, and 366 nm), blue (436 nm), green (546 nm), and yellow (577 and 579 nm) radiation. When a lamp of this kind irradiates a surface of 1 m², its effect can be calculated by specifying for each wavelength the energy W reaching the surface in a given time, or the number of photons N which, from equation (2), is the quotient of W by $h\nu$. If the energy W is given in joules and the wavelength λ in nm, we have the equation:

$$N = 5.03 \times 10^{15} \, W/\lambda \qquad (4)$$

Other sources emit a continuous spectrum, that is, all the monochromatic radiations are present in it, at least over a certain range of wavelength. This is the case with incandescent lamps, where the tungsten filament is heated to a high temperature by the passage of an electric current. It is also the case with sunlight, the chief source of light for living things on our planet.

To describe the energy distribution in a continuous spectrum, a more or less arbitrary convention must be adopted. In fact

5

since there is an infinite number of monochromatic radiations available to carry a finite amount of energy, each one can only possess zero energy. The spectrum must therefore be divided into narrow bands of finite width, and the energy contained in each calculated. Obviously the result will be influenced by the method of division. The spectrum is often divided into bands of equal width $\Delta\lambda$ in wavelength, for example 400–405 nm, 405–410, etc. But it is also usual (and perhaps better, since it has been stated that a frequency scale is more logical than a wavelength scale), to divide the spectrum into fragments of equal width $\Delta\nu$ in frequency, or, what amounts to the same thing, of equal width Δw with respect to the fundamental photon energy. These two conventions give quite different results. From equation (3) it can in fact be deduced that Δw is proportional to $\Delta\lambda/\lambda^2$. Two bands of equal width $\Delta\lambda$, situated one at the violet end of the visible spectrum $(0·4\,\mu)$ and the other at the red $(0·7\,\mu)$, would correspond, in a division by w, to bands of very unequal width, in the ratio $(7/4)^2$; say approximately 3. Consequently, if there were, for example, the same number of photons in the violet and red bands on a division by equal Δw, there would be three times as many photons in the violet band as in the red on a division into equal $\Delta\lambda$. This is no small difference. I apologize to the reader for stressing this point, but it has been the source of many errors, even among excellent authors. In this book, except in figure 2, the method of division into equal Δw will be adopted in all cases, for the reasons already given.

The *black body* is a continuous spectrum source, which plays a primordial part in the history of quanta, and is also of great importance because it is a theoretical source with which real sources can be compared. In principle the black body is an enclosed space, the wall of which is pierced by a minute aperture through which radiation is emitted. At ordinary temperatures no visible light is emitted from this small aperture, whence the expression 'dark as an oven'. If the temperature is raised, first dark red is seen, then light red, which changes to yellow, and finally to white at temperatures higher than 3000° K. The letter K, it will be recalled, denotes the *absolute* temperature, equal to the C on the more ordinary Celsius scale with the addition of about 273°.

A virtue of the black body is that the composition and energy of the radiation emitted through the aperture are independent of the material of which the wall of the oven is made. This surprising fact is explained by the reflection and diffusion of radiation from the inner surface of the wall. This occurs to a considerable degree if the area of the aperture is very small in proportion to the area of the wall. The radiation is a function only of the absolute temperature T of the oven, and Planck's famous law expresses the spectral repartition of the energy of the black body in the form:

$$N_w = \frac{c_1 \lambda^{-2}}{e^{c_2/\lambda T} - 1} \tag{5}$$

$N_w \times \Delta w$ states the number of photons of fundamental energy w, therefore of wavelength λ, related to w by equation (3), with respect to a band of equal width Δw; e, base of Napierian logarithms, and c_1 and c_2 two constants; if λ is expressed in nm, the constant c_2 has a value of $1 \cdot 438 \times 10^7$. It is worth noting that the product $N_w \times \Delta w$ is a number, therefore a quantity without dimensions; consequently the quantity N_w has the dimension of the reciprocal of one unit of energy, whence the subscript w: in actual terminology, N_w is known as the *spectral density* of the number of photons emitted by the black body, the variable characterizing the photons being their fundamental energy w. If the variable characterizing the photons were their wavelength, another spectral density N_λ would have to be defined and Planck's Law would be written:

$$N_\lambda = \frac{e' \lambda^{-4}}{e^{c_2/\lambda T} - 1} \tag{6}$$

7

As an example, figures 1 and 2 represent, for these two methods of dividing the spectrum, the spectral density of a black body at a temperature of 5,800° K. The curves are obviously different, but the areas bounded by the curves, the abscissae axis and the ordinates corresponding to the limits of the visible spectrum (0·4 and 0·7 μ) are in the same ratio on account of the relationship:

$$N_w \times \Delta w = N_\lambda \times \Delta\lambda \qquad (7)$$

These three areas are proportional to the total number of photons emitted in the ultraviolet, the visible and the infrared, that is, 4·5 per cent, 21·5 per cent and 74 per cent respectively of the total number of photons emitted by the black body at $T = 5,800°$ K. Even at this high temperature the emitted photons fall largely in the infrared; this also occurs but to a lesser extent for the energies of the three groups of photons, which represent respectively 12 per cent, 36 per cent and 52 per cent of the total of the range under consideration.

It may also be noted that the maximum spectral density depends on the convention used. The wavelength λ_m of this maximum varies with temperature according to Wien's Law, but with different constants; for the division into equal Δw and giving λ_m in nm, we have:

$$\lambda_m \times T = 9{\cdot}02 \times 10^6 \qquad (8)$$

whereas for the division into equal $\Delta\lambda$ we would have:

$$\lambda_m \times T = 3{\cdot}67 \times 10^6 \qquad (9)$$

When $T = 5,800°$ K these maxima are found at 1560 and 632 nm respectively; it is clear that the question 'where does the maximum emission from a black body at any given temperature fall?' is meaningless, since following the chosen convention it occurs in the infrared or the visible.

Real sources may have a spectral distribution similar to that of a black body. One can thus define a colour temperature T_c

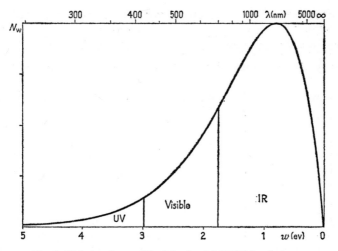

Fig. 1 Emission from a black body at 5,800°K in photons
(linear scale in fundamental photon energy *w* as abscissa)

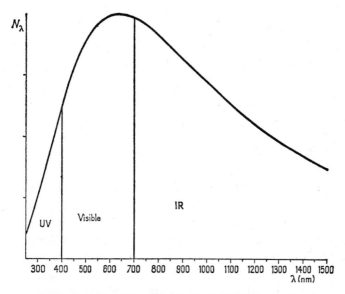

Fig. 2 Emission from a black body at 5,800°K in photons
(linear scale in wavelength λ as abscissa)

9

of the source. This is the same as that of the black body emission which falls closest to that of the source in the visible range. This brief definition is improved in colorimetry, due to colour vision. Incandescent lamps, for example, have colour temperatures

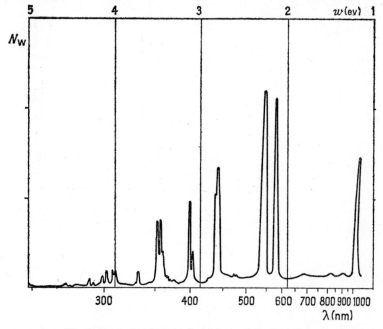

Fig. 3 Spectral emission of a high pressure mercury arc

which differ little from the true temperature of the filament (at about 3,000° K).

Other sources may emit continuous spectra which differ from that of the black body. For example, Figure 3 represents the spectrum of a mercury arc at high pressure; the lines are broader than in the low-pressure arc, and a continuous background is superimposed on the line spectrum. This source is not much used for lighting purposes because it emits almost nothing in the red. In order to remedy this, the tube is lined with a fluorescent material which, excited by the 254 nm radiation of

10

the mercury vapour (at low pressure in this case), emits a continuous spectrum on which the mercury rays are superimposed (Fig. 4). A continuous spectrum source much used in the laboratory is the high-pressure xenon arc, which gives out intense white light (Fig. 5).

The sun constitutes an important and special light source. In the absence of atmosphere, the radiation which would reach us,

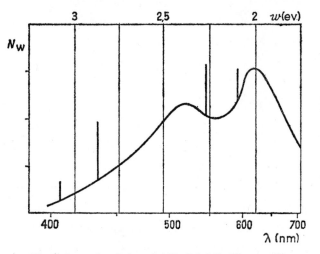

Fig. 4 Spectral emission of a 'daylight' fluorescent tube

and which can be recorded by artificial satellites, would approximate to that of the black body at about 6,200° K, but the spectra also show innumerable emission and absorption rays originating from the outer layers of the sun. In addition, both X-rays and radio-electric waves have been demonstrated in solar radiation. A surface of 1 m² placed perpendicular to the sun's rays would receive a power of about 1·4 kW, in the absence of atmosphere. Due to the equivalence between mass and energy expressed by Einstein's equation:

$$W = mc^2 \qquad (10)$$

the whole earth gains yearly a mass of about $6·8 \times 10^7$ kg from the absorption of the solar radiation. An equal mass is lost in the form of infrared long-wavelength radiation into space, assuming the earth to be in thermal equilibrium. Assuming the

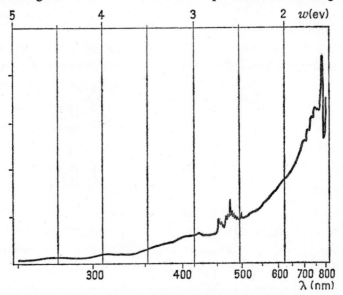

Fig. 5 Spectral emission of a xenon arc

earth to be at a mean temperature $T = 300°$ K and that it emits like a black body, we get from equation (8):

$$\lambda_m = 30\,\mu$$

Solar radiation is greatly modified by the atmosphere. Atmospheric oxygen absorbs all solar ultraviolet light of wavelengths less than 290 nm, that is about 3 per cent of the total energy emitted by the sun. Atmospheric oxygen may be present first as molecular O_2, then in the atomic state (O_2 molecules dissociate into O in the high atmosphere under the action of ultraviolet rays of wavelengths shorter than 240 nm) and finally may be present also in the triatomic state of ozone, O_3, present in the stratosphere at a variable concentration which on average would represent a layer 3·5 mm thick supposing the gas to be in a pure state under conditions of normal temperature (0° C) and pressure (1 atmosphere). In the visible range ozone has slight absorptive properties for wavelengths $\lambda > 440$ nm, which remove at most 1 per cent of the energy. In the infrared O_2 has absorption bands at 760 nm and $1\cdot27$ μ, but it is mainly water vapour and carbon dioxide that are absorbent. Water vapour can vary considerably, the height of precipitable water varying between 1 mm for a very dry atmosphere and 3 cm for a very humid one (precipitable height is defined as the thickness of the layer of water which would be formed by total condensation). Absorption also depends, of course, on the sun's elevation; if the total mass of air traversed by the sun's rays when the sun is at its zenith is taken as unity, the value becomes $1\cdot5$ when the sun's angle is 48° to the vertical and $2\cdot0$ when it is 60°.

As well as this true absorption which heats the air, there is also apparent absorption, due to scatter by both air molecules and aerosols. Molecular scatter obeys Rayleigh's Law in λ^{-4} and so explains why the sky is blue, since at equal incident energy of radiation at $0\cdot4$ μ and at $0\cdot7$ μ (the limits of visible light) the first would be scattered in the ratio $(7/4)^4$, that is $9\cdot4$ times more than the second. (The same is true for the number of photons, at an equal incident number.) The effect of aerosols varies little with the wavelength. Moreover, molecular scatter is symmetrical, and reflects back into interplanetary space half of the scattered radiation, whereas that of the aerosols is concentrated in the direction of the incident light. Finally the reflection factor of the earth interferes markedly: the sky diffuses about 35 per cent more light above ground with a reflection factor of $0\cdot25$ than it does above dark ground.

Figure 6 shows the slightly schematized appearance of the sun's spectrum at ground level, for two elevations of the sun, at 1 cm of precipitable water and 200 aerosol particles per cubic centimetre. (After Gates (1966).)

It can be seen that the idea of a colour temperature is hardly

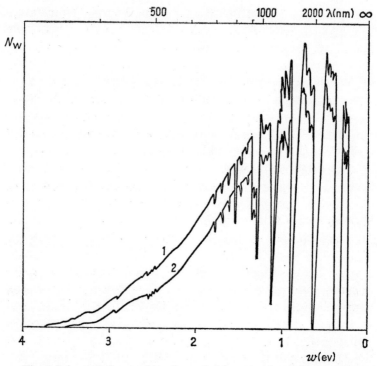

Fig. 6 Solar radiation reaching the earth (air masses 1 and 2 traversed)

applicable, but it can be approximately placed in the region of 5,800° K. Data from the same author on the total energy arriving at a horizontal surface area of 1 m² are summarized in Table 2.

TABLE II

Power radiated by the sun on 1 m² horizontal surface of the earth

air mass	total power (sun + sky)	direct rays from sun only
1·0	898 W	838 W
1·5	530 W	475 W
2·0	322 W	302 W

14

1.3. Light–matter interaction

While light passes through space, it is as though it does not exist, since it only becomes evident when it interacts with matter.

Present-day concepts of matter date from 1897, when Thompson discovered the *electron*. This negatively charged particle is one of the constituents of matter, the other being the *nucleus*. Rutherford demonstrated the existence of the nucleus studying the deflection of alpha rays from radioactive material during their passage through thin sheets of metal. While there is only one kind of negative electron, common to all chemical compounds, there is a great variety of nuclei. An atom of any given substance is characterized by its atomic number Z: if this atom is electrically neutral, its nucleus, which has a positive charge equal to Z times that of the electron, is surrounded by shells of satellite electrons. The atom can also exist in an ionized state, through the loss or gain of one or more electrons becoming either positively or negatively charged respectively.

Niels Bohr (1913) introduced the important concept that the binding energy of satellite electrons had certain definite values, and that these values constituted a definite series. The set of energy values of the Z electrons in a given state is called the *electronic configuration*. An atom can have many different electronic configurations; for example, the sodium atom (Z = 11), may have more than 120.

The electrons least strongly bound to the atom are the valency electrons, which are responsible for chemical reactions. Their bond energy varies between 3 and 20 eV. For ionization to occur by the displacement of an electron from the atom the amount of external energy applied must exceed the bond energy.

One photon of visible light cannot therefore cause ionization, and even the 254 nm ultraviolet mercury ray can only, according

15

to equation (3), ionize atoms whose bond potential for the valency electrons is less than 4·9 V. In fact only X-rays and gamma rays are true ionizing rays.

Ionization phenomena will not be considered in depth in this volume. This is not because they are without effect on living matter; on the contrary, ionization results in important biological changes. In the first place, the positive ion has chemical properties quite different from those of the neutral atom from which it arose, since its valency differs by one unit. Secondly, if the ejected electron attains sufficient speed, it also can cause ionization over the length of its track, and this secondary ionization is sometimes important. At the end of its track, it becomes attached to a neutral atom giving a negative ion, which rapidly becomes hydrated with the liberation of heat. The reason for not dealing with ionization is simply because it constitutes a separate aspect of biology known as *radiophysiology* which is distinct from *photophysiology*, with which we are concerned in this book.

It should also be noted that ionization can result from changes other than the arrival of a photon in an atom. For example a rapidly travelling charged particle (e.g. alpha or beta rays from radioactive material) can also displace an electron as a result of the reaction of the electric field of the incident particle on the charge of the electron. In this ionization the particle loses an amount of energy equal to that which it gives to the electron, that is to say, the energy of the bond plus the kinetic energy acquired by the electron leaving the atom. Following this exchange the particle will continue in its path at a lower speed. The ionizing photon, on the other hand, usually loses all its energy to the electron and disappears. In fact it is impossible to slow down the photon because its mass at rest is zero, it only exists in the condition of moving at a speed c. However, a process is known called the *Compton effect*, in which the recoil collision of a photon and an electron results in a reduction of the energy of

16

the photon, whose wavelength is thus increased. (From the wave point of view the change in wavelength can be interpreted as a Doppler-Fizeau effect due to the scattering of light by a moving object.)

An uncharged particle, like the neutron, does not directly cause ionization, but its recoil collisions with nuclei, in particular the nuclei of hydrogen (protons) which are abundant in living matter, turn these into secondary charged particles which are capable of causing ionization.

Leaving aside ionization, we will now pass on to consider other effects of radiation on matter, grouped under the general heading of *excitation*. The energy lost by the photon to the atom does not eject an electron, but it modifies the electronic configuration: one of the satellite electrons passes from the *fundamental level*, which has the lowest energy, to an *excited* higher-energy state. According to Bohr's concept the electron can be considered to pass from its original orbit to an orbit which is farther away from the nucleus, but remains nevertheless bound to it. Excitation requires less energy than ionization, and it is the only effect that visible and ultraviolet radiation (at least the sun's ultraviolet) can produce.

The study of excitation reactions following the effects of photons on atoms and polyatomic molecules and secondary phenomena following excitation constitutes *photochemistry*. Photochemistry has specific characteristics, compared with ordinary chemistry. In the latter, thermal energy is sufficient to start off reactions, but this thermal energy is weak, since at 20° C the average energy of translation of a molecule is only 0·04 eV. On the other hand, one photon of blue light carries 60 times this energy, and the reactions which it can set off would only be thermally possible at about 18,000° C.

Another difference is that a photon only excites through being absorbed, it disappears and all its energy is transferred to a single molecule, which it excites. With the usual radiation intensities, the proportion of excited molecules remains small.

17

In changing electron-volts into joules, equation (3) is written as follows, expressing λ as before in nm

$$w = \frac{1,987}{\lambda} \times 10^{-19} \text{ (joules)} \qquad (11)$$

Monochromatic radiation of 1 Watt transports $1/w$ photons per second, say for example $2 \cdot 5 \times 10^{18}$ photons if $\lambda = 500$ nm; if this radiation is absorbed by a gram molecule of matter, which contains $6 \cdot 025 \times 10^{23}$ molecules, only one molecule in about 240,000 will be excited per second; if the very short life of excited molecules is taken into acount, the proportion of excited molecules at a given instant is decreased further by a factor 10^{8}.

Because of the very small proportion of excited molecules surrounded by the many normal molecules, an excited molecule in practice only reacts with normal molecules. The laws of photochemistry will be *linear* in consequence. Non-linear effects are revealed only with very intense light flashes.

In photobiology, it is essentially polyatomic molecules which react with light photons. These molecules are complicated arrangements of nuclei and electrons. Some of these electrons constitute the electron shells of specific nuclei, but the photons do not affect them because their bond energy is too high. Others, which are analogous to the valency electrons of atoms, do not belong to any definite atom but to quasi-autonomous groups of atoms in the molecule, the *radicals*. The energies which separate the levels of these electrons are of the order of an electron volt, and consequently photons of visible and especially ultraviolet radiation can induce electronic transitions. In addition, in each of these electronic configurations the molecules and radicals may have different possible states of vibration and rotation. These also have different energies but are closer to one another than the electron levels: the separations between them are of the order of $0 \cdot 1$ eV, and are thus in the infrared range. The number

18

of energy levels of a large molecule, such as those which constitute living matter, is extremely great because of the numerous possibilities of reciprocal vibrations in the atoms and the groups of atoms in the molecule, and the mutual interactions between vibration, rotation and electronic configurations. Looked at in this way, the complexity of photobiology can be understood.

1.4. Principal photochemical effects

When a photon passes through a molecule there is no interaction unless the photon energy is equal to the difference between two energy levels in the molecule. *A priori* it would seem that this strict equality would mean that no photon of a complex incident radiation would have exactly the energy required; fortunately the *uncertainty principle* broadens the levels slightly. The shorter the time a molecule remains in a given electronic configuration the more marked this will be.

If the photon finds no suitable energy difference in the molecule, it continues in its path without energy exchange with the surrounding matter. During its passage the molecule will be distorted by the electric field of the wave, this being the origin of dispersion and the refractive index. Very approximately the duration of this changed state of the molecule is of the order of size of the inverse of the frequency v, that is in the photobiology field about 10^{15} seconds.

If on the other hand there is an interchange of energy between the photon and the molecule, the photon disappears (absorption), and the molecule becomes excited. This is the *primary* photochemical reaction, and it is practically independent of temperature. This fact can sometimes be used to separate it from secondary reactions occurring in the dark, which are markedly temperature-dependent and, for example, very much slowed in liquid nitrogen. In 1918, Stark first successfully distinguished these two successive phases of photochemical reactions. While the fundamental state of minimum energy is stable, since it needs a supply of energy to be changed, an excited state is *metastable*, tending to fall back spontaneously into the fundamental state, losing its excess of energy in some way. Theoreticians think that if the molecule were isolated this metastable state might last indefinitely, but the disturbances by collisions with neighbouring molecules, traces of impurities, electric or magnetic fields, mean in fact that the metastable state has an average duration of the order of 10^{-8} seconds.

In photobiology, considering the weak energy of the photons used, only one excited electronic configuration is possible, that which differs least in energy from the fundamental state. In some cases an excited *triplet* may exist, with energy intermediate between the fundamental state and the first state of ordinary excitation. It applies to molecules with two valency electrons with the same energy in the fundamental state: they are antiparallel, that is, their *spins* (moments of self-rotation linked to a magnetic moment of the electron) are in opposite directions. In the excited triplet state, one of these electrons passes to an energy state different from the other; they continue meanwhile to act on each other but with their spins parallel and in the same direction. These excited triplet states are of markedly longer duration than the normal excited states (10^{-4} second to 1 second). For this reason a triplet phase is often mentioned in photochemistry.

What happens when the metastable state ends? A great variety of effects is then possible. The simplest is direct optical *dissociation*. The absorption of a photon alters the molecule to an excited state where the energy of vibration becomes greater than the bond between the radicals, and the molecule splits in two. It can also produce an indirect dissociation by a more complicated mechanism; to understand it the energy levels of a molecule will be schematized graphically (Fig. 7). N, F and M represent three different electronic configurations, each symbolized by a pyramid indicating different states of vibration and rotation. The lines represent the energy levels, which increase with the height of the line. The absorption of a photon causes transition 1 from N to F, then an *internal conversion*, 2, causes the molecule to change from a state of high electronic energy and low oscillatory energy to a state of minimum electronic energy (fundamental state) but with a strong vibrational energy which produces dissociation; this will be partial

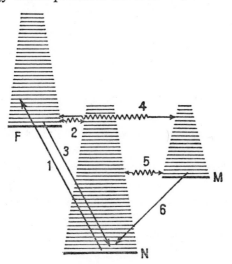

Fig. 7 Photochemical processes (energy diagram)

since it only acts on the molecules which have undergone the the internal conversion.

Another possibility is *fluorescence*; the incident photon causes transition 1 and during the duration of the metastable state (about 10^{-8} sec.) the excited molecule loses a little vibratory energy through collisions. Transition 3 represents the return to the fundamental level, with the emission of a photon of wavelength slightly longer than that of the exciting photon due to the above-mentioned loss. In theory, the outcome would be unity, with a photon lost for each one absorbed, but this is not in fact the case because of the internal conversions and also as a result

21

of an unexplained migration of energy, when the molecular concentration is high. It is also generally true that photochemical reactions decline paradoxically when the molecular concentration increases beyond a thousandth of normality.

Phosphorescence is a somewhat different process; here there is an internal conversion to a metastable state M (transition 4) of longer duration and a subsequent return (5) in the opposite direction is followed by the emission of photons. At low temperature transition 6, which gives 'long-life fluorescence', is also observed.

Another possibility is for the excited molecule to react chemically with a neighbouring normal molecule. The excited valency electron is situated (using the terminology of orbits) in an orbit more weakly linked to the molecule, and one which is therefore stretched over a greater volume than the base orbit. For this reason it reacts more easily with a molecule short of electrons. Absorption of light has caused the molecule to become a *reducing agent*. It may also be said that the excited electron is no longer paired with the electron remaining in the fundamental state, and can form a paired-electron bond with a single electron from another molecule, which results in a stable substance. Sometimes the excited molecule gives up a proton, when its bond with the molecule is weakened by removal of the excited electron. In this case there is a pH change and the excited molecule becomes a weak acid. It is even possible that the excited 'hot' molecule expels an H_2 group and forms two new stable molecules, by a process analogous to the dissociation following internal conversion.

In a crystalline medium, the action of light can cause *photoconduction*. This is particularly the case in ionized crystals, for example photographic silver bromide crystals. The crystal is a regular lattice of Ag^+ and Br^- ions. The absorption of a photon makes one electron of a Br^- ion pass to the excited state and instead of staying bound to that ion, it passes in a 'conduction band', that is, it can move freely within the crystal. An external electric field will set this electron in motion, and an electric current is thus generated; the crystal has become a conductor although it is an insulator in the dark. The photon, in removing an electron from the bromide ion, transforms it into a 'positive hole' in the regular crystal lattice. There are thus several possibilities: the conducting electron can recombine with the positive hole, giving the initial Br^- ion (an analogous process to the return from the excited to the fundamental state, with the possible emission of radiation). It can also neutralize another positive hole, which means in effect that, like conducting electrons, positive holes can shift within the crystal. Lastly the electron can neutralize an Ag^+ ion, giving a neutral Ag atom. This last process forms the latent photographic image; the first reaction is unwanted and has therefore to be prevented: this is achieved because the impurities in the crystal produce distortion of the ionic lattice. In particular, traces of sulphur in the gelatin form Ag_2S deposits at the surface of the crystals, and it is on contact with these that the neutral Ag atoms form the latent image. It is not known whether the impurities fix the conducting electrons or the positive holes after migration. The grains of the photographic emulsion are triangular or hexagonal crystals, with sides of about 1 μ, and containing 10^9 Ag^+ ions. The latent image consists of small groups of 3 or 4 atoms at the crystal surface and these groups catalyse the action of the reducing bath during development, until the entire crystal is transformed into metallic silver. The catalytic action of the latent image during development results in an amplification of about 10^8, which is more than that of high-gain photoelectric cells used in television. Although the discussion of photography may appear far removed from photobiology, it is in fact relevant since some proteins can act as semi-conductors, and photo-conductivity thus has a biological role.

The final possibility resulting from the excitation of a molecule by a photon is that nothing happens. In this case all the energy of the absorbed photon is dispersed into the

23

surrounding medium as heat. This is the commonest reaction in solutions because the effects of collision between active and solvent molecules are considerable. For example if the isolated molecule becomes dissociated, the 'cage' effect due to the barrier of solvent molecules facilitates collisions between the dissociated fragments and helps in their recombination, especially if the incident photon has enough energy to transfer considerable kinetic energy of agitation to the fragments. Similarly, after an internal conversion, collisions hinder the inverse return through loss of vibratory energy, and the excess of energy is degraded into heat. Fluorescence, also, is often blocked by the solvent or by impurities. Lastly, it may be simply that the excited molecule becomes very 'hot', that is, has a high oscillation energy with respect to the average thermal energy. In this case there will be an exchange to the profit of the solvent molecules with which the excited molecule is in a permanent state of collision: at room temperature, each water molecule of the solvent undergoes about 10^{12} impacts per second. The active molecule in solution will receive more because it is much larger and so, while the excited state lasts, say for 10^{-8} seconds, the number of collisions with the solvent must be estimated in the range of hundreds of thousands at least. Even if the product were something other than heat the yield would be mediocre when the active substance was in solution.

REFERENCES

GATES, D. M., Science, **151,** 523 (1966)
SELIGER, H. H. and MCELROY, W. D., *Light: Physical and Biological Action*, N.Y. and London, Academic Press (1965)

CHAPTER II

Techniques in Photobiology

2.1. Photochemical laws

Although the action of light on matter has been observed since earliest times, the first scientific photochemical research may be attributed to Scheele (1742–1786), who studied the blackening of silver salts in the violet end of the sun's spectrum. In 1785 Berthollet observed the effect of sunlight on an aqueous solution of chlorine, and de Saussure (1796) designed, using this principle, the first *actinometer* for measuring light intensity.

The early work of Niepce (1814) on photography stimulated the development of photochemistry. Grotthus (1818) stated the first photochemical law: only the radiation absorbed by a substance produces a photochemical effect. This law was confirmed by measurements made by Draper (1839) and often carries the names of both workers. Today in the light of modern ideas on energy it seems almost a truism. The converse is not true: the absorption of radiation by no means always produces a measurable photochemical effect and, as we saw at the end of the last chapter, the only tangible result of absorption is in most instances a heating effect on the substance.

The second law of photochemistry was enunciated by Bunsen (1811–1899) and Roscoe, in relation to their study of the combination of chlorine with hydrogen under the action of light. This law, the law of *reciprocity*, states that photochemical action depends only on the product of light intensity and the duration of exposure. In present-day terminology, the ambiguous term, intensity, is better replaced by *energy flux*, which describes the

25

power transported in the form of radiation. If the flux is denoted by Φ and duration of exposure by t, the quantity:

$$W = \Phi \times t \qquad (12)$$

is the radiant energy absorbed if the flux remains constant during the whole time of exposure. This law of reciprocity only holds for primary photochemical action and cannot be applied to secondary reactions. In sunstroke, for example, the result is very different if the duration of exposure t is multiplied by 100, by dividing by the same amount the energy flux of the ultra-violet radiation received by a given epidermal surface.

When the Quantum theory was introduced at the beginning of the twentieth century, photochemistry acquired a firm theoretical basis. Stark and Einstein's law of *photochemical equivalence* (1908), states that each atom, ion or molecule which undergoes a primary reaction absorbs a quantum of light, or photon. Of course this simple explanation will be modified by the fate of the excited molecule, especially if subsequent secondary reactions develop. Thus the *quantum efficiency Q* of the overall photochemical reaction can be defined as the ratio of the number of molecules which finally undergo the reaction, to the number of photons absorbed in the primary reaction. The efficiency is generally not high and this point will be dealt with frequently in this book. The following figures are quoted as examples. The value of Q for assimilation by chlorophyll is 25 per cent at most in the laboratory, and reaches barely 2 per cent under natural conditions making use of solar radiation. The same value (0·02) characterizes vision (at the absolute threshold) and the inactivation of phage S-13 by ultraviolet light. With complex radiation, efficiency is often decreased by the phenomenon of *reactivation*, since a light of longer wavelength than that which caused the photochemical reaction can partially destroy the effect. Sometimes, on the contrary, a *chain reaction* is set off by the action of light. The excitation passes like an infection from one molecule to another, so that millions of molecules react (obviously only possible for an exothermic reaction). For example, the quantum efficiency of the photochemical combination of hydrogen with chlorine, resulting in the excitation of the latter molecule, is of the order of 10^6.

The quantum efficiency can be considerably augmented by association with active molecules of some coloured substances. Vogel (1873) made the chance observation that a photographic emulsion which is relatively insensitive to green light becomes much more so after immersion in a bath of yellow dye (which he was testing for anti-halo effect). This sensitization has been extended into the red and even into the infrared (up to $\lambda = 1\cdot2\ \mu$) by dyes incorporated into the gelatin. The maximum effect is obtained with a monomolecular layer of dye on the surface of silver bromide crystals. Sensitization can be explained either by a transfer of energy from the dye to a Br^- ion after absorption of a photon, or by a transfer to the crystal of an electron freed by the dye.

This photosensitization by dyes can be utilized in biology. Raab (1898) found that Protozoa in solution coloured by acridine, which in itself had no action on the cells, were killed if they were irradiated with a normally inoffensive visible diffuse radiation. These important phenomena will be considered again later on.

One last point which must be made with regard to the laws of photochemistry concerns the symmetry of light. Natural light usually represents a symmetry of revolution about each light ray, but light can be *polarized*, that is, given another symmetry. For example, in light polarized rectilinearly, the electric and magnetic fields remain in fixed directions (perpendicular to one another) in the plane of the wave. If a molecule is excited by light of this kind, the radiation emitted as fluorescence is in theory itself polarized, but interactions with other molecules during the state of excitation (10^{-8} s.) cause the partial or total loss of this light-polarization 'memory'. But if the molecule possesses a natural symmetry instead of being isotropic, its orientation can be modified by polarization and may so persist for quite long periods of time, thus explaining the residual persistence of polarization sometimes observed in phosphorescence.

Another possible form of light symmetry is *circular* light, which may revolve to right or left. A similar form of symmetry is present in many organic molecules which possess a rotatory power to right or left. This means that the direction of the electric field of the rectilinearly polarized wave undergoes a deviation in one or other direction in its transit through the substance. Chemical reactions carried out in the laboratory always give racemic mixtures of left- and right-handed varieties of molecules in equal proportion. On the contrary sugars and amino acids produced by organisms are optically active, this optical symmetry being characteristic of life. From the thermodynamic point of view the mixing of two active varieties to form a racemic mixture liberates energy. An optically active state is not as stable as a racemic one, and, in one sense, constitutes an excited state. The question arises as to whether naturally occurring asymmetry is photochemical in origin. The presence of an excess of light circularly polarized to the left over that polarized to the right in natural light could in fact explain its origin.

One of the currently favoured hypotheses on the origin of life postulates that macromolecules were initially produced by photochemical polymerization in a hydrocarbon 'soup' (estuary or marine shore) under the action of short-wave ultraviolet light which was then able to reach the earth's surface from the sun because the atmosphere contained no oxygen. Near its absorption band at about 185 nm, atmospheric water vapour in the earth's magnetic field shows a slightly different absorption for light polarized to left than to right. Light reaching the surface of the primitive soup would therefore have a slight excess of light circularly polarized to the left in the northern hemisphere, and to the right in the southern hemisphere. An attempt to verify this theory has been made by measuring the residual optical activity of Precambrian sediments, but unfortunately it seems that the magnetic field of the earth has been reversed several times during geological time. It seems likely, however, that photobiology has probably played a part in the development of the natural asymmetry of living forms.

2.2. Spectrophotometry

The quantitative study of a photochemical reaction can be made by optical means, or by chemical analytic methods.

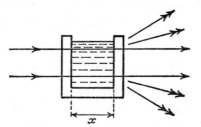

Fig. 8 Spectrophotometric container

Firstly, the basis of the spectrophotometric method will be outlined.

Let us consider a container with parallel sides, preferably made of quartz, so as to be transparent to ultraviolet light and containing a thickness x of a solution of substance A which is to be studied (Fig. 8). A cylindrical beam of light of wavelength λ carrying energy flux Φ_0 is shone on to the container at right angles. After passing through the container the beam carries a smaller flux Φ and the *transmission factor* of the container for the radiation under consideration is defined by the ratio, always less than unity:

$$\tau_\lambda = \Phi/\Phi_0 \qquad (13)$$

The difference between the incident flux Φ_0 and the transmitted flux Φ is not entirely absorbed into the solution, a small fraction is reflected by the sides of the container and returns in the initial direction of the beam. This can be eliminated by using the symbol Φ_0 to denote not the incident flux, but the flux which would be transmitted by the same container full of a pure solvent (supposedly transparent; possible auto-absorption is eliminated at the same time by this method). If the solution is not perfectly transparent light is scattered outside the normally transmitted beam. This is indicated by the double-headed arrows (Fig. 8). This difficulty, which will be dealt with later, is often encountered, since biological solutions tend to be turbid.

Instead of the transmission factor, it is preferable to use the *optical density* defined as the decimal logarithm of the reciprocal of the transmission factor

$$D_\lambda = \log_{10} 1/\tau_\lambda = -\log_{10}\tau_\lambda \qquad (14)$$

A perfectly transparent medium has an optical density of zero. A density of 1 means that 10 per cent of the flux is transmitted, a density of 2, 1 per cent, etc. The concept of optical density, which might at first seem curious, has many advantages. First of all, the *absorption factor* $1 - \tau_\lambda$ remains proportional to D_λ as long as the density is low; taking into account that $\log_{10} e = 0.43$, we have;

$$1 - \tau_\lambda = 1 - \exp(-D_\lambda/0.43) = \frac{D_\lambda}{0.43} - \frac{D_\lambda^2}{2(0.43)^2} + \cdots \quad (15)$$

Further, if the substance A is alone in solution, and undergoes no chemical change under the action of the radiation λ, the absorption obeys *Beer's law* (1855):

$$D_\lambda = 0.43 \, \alpha_\lambda \times C \times x \qquad (16)$$

where C denotes the concentration of substance A (mass dissolved in unit mass of solvent) and α_λ is the *extinction coefficient* characteristic of the dissolved substance, independently of its concentration and of the thickness of the container (the dimensions of α_λ are the reciprocal of length).

A third advantage of density is that, if several substances in a cell of given thickness x have the densities D_1 for concentration C_1, D_2 for concentration C_2, etc. (measured separately), a mixture of these substances, if they do not react chemically with one another, will have an optical density $D_1 + D_2 + \ldots$ at the same thickness x and same concentrations.

Let us return to the case where a single substance A, present in the container, is to be studied in the range of wavelengths where there is no photochemical reaction. The only effect of measurement will be to heat the solution slightly. In this case substance A will be optically characterized by its *absorption spectrum* obtained employing λ or w as abscissa (the latter is used here) and the coefficient of extinction which is obviously a function of λ or of w, as ordinate. α_λ is usually only known in relative value since the concentration is often unknown. An elegant way to get over this difficulty is to take as ordinate the ratio D_λ/D_m, where D_m represents the maximum optical density of the solution over the range of wavelengths studied. The maximum occurs at a certain wavelength λ_m. From (16) it follows that:

$$D_\lambda/D_m = \alpha_\lambda/\alpha_m \tag{17}$$

and the resulting curve independent of C and x is characteristic of substance A.

Contrary to what we have found for energy in a continuous spectrum, the maxima of which occur at 4 different radiations according to whether λ or w is used for the abscissa, and photon number or spectral energy density as ordinate, in this case there is only one perfectly defined maximum: from (13) the value of τ_λ is the same whether the flux is measured as energy or as number of photons (per second), since it is a simple ratio. The minimum τ_λ, and therefore the maximum D_λ, is consequently obtained for the same radiation, whether calculated in λ or in w. The position of maximum absorption is an absolute value, whereas that of the maximum of a continuous emission spectrum is not.

All the absorption curves found in this work are plotted with a linear scale in w as abscissa (fundamental energy of the photons used to measure absorption) and D_λ as ordinate, taking $D_m = 1$ for simplicity. From (17), these curves also give the

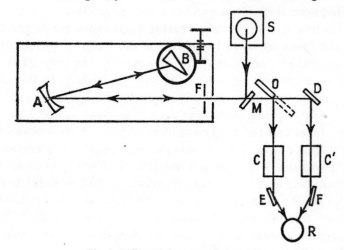

Fig. 9 Differential spectrophotometer

values of the extinction coefficient relative to the maximum extinction coefficient α_m.

Many spectrophotometers are commercially available. The principle of only one will be described. This allows direct measurement to be made of the *difference spectrum* between samples contained in identical containers C and C' (Fig. 9). Light from the source S is reflected from a mirror M through the slit F of the monochromator, in front of a concave mirror A. The light is dispersed by a prism B with the posterior surface silvered. A spectrum is formed in the plane of the slit F which is now the exit aperture, and the desired monochromatic radiation is isolated by rotation of B. The reflecting face of B is not exactly

32

perpendicular to the plane of the figure; because of this the light, in leaving the monochromator, passes a little above the mirror M, is reflected on an oscillating mirror O and the fixed mirrors D, E, F, and reaches the photoelectric receptor R after passing through one or other container. The receptor is used over a range where its response is a linear function of the flux which it receives, and the apparatus registers directly the ratio of currents resulting from fluxes Φ and Φ' which have passed through the containers. The decimal logarithm of this ratio $(\tau_\lambda/\tau'_\lambda)$ gives the difference $D'_\lambda - D_\lambda$. If the container C contains pure solvent and C' the substance to be measured, the desired absorption spectrum is directly obtained, and inaccuracy due to reflection in the container and auto-absorption in the solvent are eliminated.

As has been previously mentioned biological suspensions often scatter light and this is an important cause of error. The scatter can be decreased by selecting a solvent with a refractive index close to that of the particles in suspension. A more satisfactory method is to catch all the light leaving the container, including both the transmitted beam and the scattered light. To achieve this, an opalescent glass is placed against the wall of the container through which the exit beam passes and the photo-receptor is placed in contact with this glass. In principle a more satisfactory method is that using *Ulbrich's sphere*, which is a hollow sphere coated inside with magnesium oxide. One of two small windows in the wall of the sphere collects all the light emerging from the container. The other (placed outside the transmitted beam) allows a flux, proportional to the total flux leaving the cell, to be sent to the receptor.

The absorption spectrum characteristic of a given substance allows the presence of the substance to be detected; when this spectrum contains absorption bands which overlap, it is often useful to determine the curve giving the first derivative of D_λ as a function of λ (or w) and which in this case is easier to recognize than the absorption curve itself. Apparatus can be constructed to give this derived curve directly. French (1959) has routinely employed this method to study the structure of the red absorption band of chlorophyll *in situ*.

If the solution contains a mixture of substances which do not react with each other, their densities can be added, and if the coefficients of extinction of the suspected constituents of the mixture are known, these can be quantitatively analysed by solving a system of linear equations. Special electronic calculators have been constructed to operate on this principle on the results given directly by the spectrophotometer.

2.3. Action spectra

The absorption spectrum of a substance A gives the relative probability of photons being absorbed by it as a function of their fundamental energy w, but it gives, as we have already said, no information about the possible photochemical action which may result from the absorption. For example, in the infrared a photon does not cause excitation, although it can be absorbed by the molecule. Since its energy w is too weak to alter the electronic configuration, it can only increase its vibrational and rotational energy, and therefore the temperature, and when the latter returns to normal after collision with other molecules, the net result of the reaction is that heat is given to the solution. In the visible and ultraviolet range, a photon can cause excitation, but frequently (especially in solution), the photon energy is only transformed into heat. From the first law of photochemistry, a photon must be absorbed in order to cause excitation, and the absorption spectrum indicates whether it will be absorbed or not. This, however, provides inadequate information and therefore the photochemical action of photons as a function of their fundamental energy w, that is, the *photo-sensitivity* as a function of w, is measured in terms of the *action spectrum*.

34

The direct measurement of an action spectrum is much more difficult than that of an absorption spectrum, and makes several further assumptions necessary. Suppose that photochemical action transforms substance A into substance B, and that the transformation can be followed by some physical or chemical method. The simplest way to do this is by measuring the absorption in the mixture of a radiation (red or infrared, for example), the quantum energy (w) of which is too small to cause photochemical excitation. Usually α_λ will be different for A and B and it is thus possible to follow the progressive transformation of A into B under the action of a photochemically active radiation of shorter wavelength. Taking N_w as the number of photons of quantum energy w, and ε_w the probability that a photon will be effectively absorbed by a molecule, or fraction of a molecule of A which is converted to B and taking Q the quantum efficiency, then the number of molecules of A converted is proportional to the product $N_w \varepsilon_w Q$ if the change in concentration of A remains small throughout the experiment. If photons of different quantum energy w' are used and the process is stopped when the same amount of A has been converted to B as in the first experiment, then the number of molecules of A which have been transformed will be the same, thus we have:

$$N'_w \varepsilon'_w Q' = N_w \varepsilon_w Q$$

In general the quantum efficiency Q is constant, whence:

$$\frac{\varepsilon'_w}{\varepsilon_w} = \frac{N_w}{N'_w} \tag{18}$$

By this method the photosensitivity ε_w can be measured as a relative value, and an action spectrum constructed taking w as abscissa, and as ordinate the ratio $\varepsilon_w/\varepsilon_m$ (ε_m indicating the maximum photosensitivity, similarly to α_m). Unlike α_λ, the

35

curve ε_w depends on the choice of the number of photons or their total energy, but not on the choice of abscissa. This direct measurement of the action spectrum is rather difficult. Stable monochromatic radiation is essential, and this is usually obtained using a continuous spectrum source and a mono-chromator, or narrow-band interference filters. The energy (from which the number of photons is calculated) is measured with a calibrated thermopile, but several precautions are necessary (including protection of the receptor against infrared radiation emitted by the experimenter). Photoelectric receptors are more sensitive, but they need frequent calibration against a thermopile. One of the basic assumptions of equation (18) relates to the secondary reactions occurring between A and B. Either of the latter may become saturated above a certain number of photons absorbed per second. It is important to be certain that the photochemical action is really light-controlled. When several photosensitive substances are present at the same time, the simple addition of densities which applies for absorp-tion spectra is frequently not valid for action spectra. This is because energy transfers between an excited molecule of one kind and a normal molecule of another are frequent. Chloro-phyll photosynthesis is a good example of such a complex interaction.

Fortunately, in some cases, the measurement of absorption spectra gives information about the photosensitivity. For example, if the absorption due to the initial product A, and that due to the *photoproduct* B which is formed on disappearance of A are in different regions of the spectrum, the photosensitivity can be determined from the spectral difference between A and B. In addition impurities are eliminated by this method. The study of visual pigments provides a good example of this principle.

Another very different technique is to use an extremely intense and brief source or 'flash' obtained by the discharge of condensers into a tube filled with gas at low pressure. The duration of the flash is 10 μ s or even less, and the absorption spectrum is recorded photographically. In many cases, the dark secondary reactions do not have time to begin during this time and the primary reaction itself can be observed. The laser, can also provide a source of brief intense monochromatic light. One inconvenience of this technique is that the concentration of activated molecules can become high enough for them to interact, yielding two normal molecules per collision. This results in, on one hand, a drop in the quantum efficiency, and on the other, deviation from the usual linear photochemical laws. One advantage, however, is the appearance of free radicals which do not have time to react during the flash, which means that the initial intermediary reactions of a photochemical reaction can be identified. This method is being further developed.

2.4. Electronic resonance method

Besides the optical techniques which we have just considered, analytical chemical methods can, of course, be useful in the study of photochemical processes. The short life of some of the intermediary substances and their presence in only trace amounts means that the methods must be precise and are difficult to carry out. These will not be considered in detail and the principle of only one physico-chemical method will be explained. This method is based on the magnetism of the atom, and is known as Electron Spin Resonance (ESR).

As we have said before, each satellite electron in an atom rotates on itself; this is known as spin, and from this a permanent magnetic moment is set up and each electron can therefore be regarded as a small magnet. According to Pauli's principle, not more than two electrons in the same energy state can exist in an atom or molecule. If there are two, their spins are opposed and their magnetic moments therefore neutralize each other. On the other hand, atoms or molecules with an uneven number of electrons have of necessity an unpaired electron. This results in a magnetic moment and they are therefore paramagnetic. When placed in a magnetic field, they tend to become oriented so that the moment is parallel to the field, but this orientation is upset by thermal agitation, and the total magnetic moment will be at a first approximation inversely proportional to the absolute temperature T.

Energy conditions favour the pairing of electrons, so that atoms with an uneven number of electrons combine to form diatomic molecules with an even total number of electrons (H_2, N_2, Cl_2). Alternatively they may, like the alkali metals, form positive ions with an even number of electrons. In the solid state the valency electrons become paired in the conduction bands.

Organic molecules are not usually paramagnetic: but the breaking of a covalent bond by a photon (*photolysis*) yields two radicals of which each has an unpaired electron and will therefore be paramagnetic. But these radicals are usually short lived (except for certain aromatic free radicals) unless they take part in polymerization reactions. Triplet states, which have been previously mentioned in connection with molecular excitation induced by light, have two unpaired electrons and are therefore paramagnetic. The normal state of the oxygen molecule O_2 is also a triplet state, so that it is strongly paramagnetic. The first excited state of O_2, called *singulet*, seems to play a very important role in photobiology.

The principle of ESR technique is as follows: a uniform magnetic field H_0, which is adjustable up to approximately 5,000 oersted, is applied to the substance to be analysed. In the absence of a field, the magnetic moments of paramagnetic atoms are randomly oriented; the field tends to orient them, but the torque thus produced behaves like the weight on a gyroscope, because of the rotational motion associated with spin. The moment takes up a precession about the direction of the field H_0, and describes a cone (Fig. 10) with a frequency given in megahertz by the equation:

$$\nu = 2 \cdot 8\, H_0 \qquad (19)$$

A second magnetic field H_1 is applied perpendicular to H_0 and varies sinusoidally with time. If the frequency of H_1 is equal to the frequency of precession ν, there is a transfer of energy by resonance between the source of energy which creates the field H_1 and the system of moments in precession. The latter then reverse their direction to one opposite to H_0. The detection of this resonance, by sudden variation of the

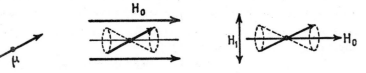

Fig. 10 Electronic spin resonance

energy supplied to create the alternating field H_1, is the basis of ESR technique. In practice it is usual to work at about 9,500 MHz and to vary H_0 until resonance is obtained. Unfortunately, if the sample contains water, the dielectric loss caused by the agitation of polar water molecules greatly weakens the acuity of resonance. This can be remedied by working at low temperatures since the loss in ice is less than in water. The sample can also be placed in a cavity resonator at the nodal point of an electric field.

The appearance of resonance indicates the existence of unpaired electrons in the sample, and the intensity of resonance gives a measure of the number of these electrons, since a comparison is made with a standard sample where the number of electrons is known. Further information is obtained from the *hyperfine structure* of the ESR spectrum, resulting from interaction with the magnetic moments of protons in the vicinity of the electron under consideration. It is known that atomic nuclei also have moments which are used in the nuclear resonance technique which has become so important in contemporary physics. The moment of a proton is about 2,000 times less than that of the electron, and the resonance ray is therefore replaced by several rays very close together. From this, important conclusions can be drawn about the structure of the molecule being studied.

The application of ESR in photobiology is complicated by an experimental difficulty. Samples are usually sectioned and manipulated with steel microtomes and forceps, which may contaminate the preparation with a microscopic fragment of iron alloy. The resulting ferromagnetism (enormous compared with the paramagnetism of the sample) will produce wide parasite resonance bands independent of temperature, whereas the paramagnetic rays vary with 1/T.

ESR technique has up to now been chiefly used in photobiology for studying photosynthesis. This is facilitated by the presence of a characteristic signal always shown by chlorophyll, which is independent of temperature. This suggests that it corresponds to the primary and purely physical process of photosynthesis, perhaps linked to a semi-conduction with immediate separation of charges preventing the recombination of electrons. ESR is, however, being more and more widely used to study the effects of ultraviolet and visible light on various organic animal substances.

REFERENCES

BOWEN, E. J., ed., *Recent Progress in Photobiology*, Oxford, Blackwell (1964)

CALVERT, J. C. and PITTS, J. N., *Photochemistry*, N.Y., Wiley (1966)

FRENCH, C. S., Proc. 19th Ann. Biol. Colloq. Oregon State Coll., 52 (1959)

GIESE, C., ed. *Photophysiology*, 2 vol., N.Y. and London, Academic Press (1964)

McELROY, W. D., and GLASS, B., ed., *Light and Life*, Baltimore, Hopkins (1961)

CHAPTER III

Ultraviolet and the Living Cell

3.1. Harmful effects of radiation

In the last two chapters, we have discussed the physical basis of the action of light on living matter, and we will now go on to our own particular subject. The action of light on plants will not be considered, as this would warrant a whole volume to itself, and we shall be solely concerned with the action of light on animals. We shall begin at the bottom of the scale by examining the photobiology of unicellular animals and of cells isolated or in suspension.

Ultraviolet is the essential factor in this action, and for the sake of brevity we shall refer to it as UV. Its harmful and even lethal effects have long been known. In the case of ionizing radiation, harmful effects are due to toxic agents (peroxides) produced by radiation in the surrounding aqueous medium. UV can also produce peroxides or other toxic agents, but these substances are produced in the cell itself, as a result of absorption by macromolecules, such as nucleic acids, which absorb specifically at about 260 nm, and proteins which absorb at about 280 nm. The mass of the cell, or *cytoplasm*, contains most of the protein and its *nucleus* contains both nucleic acids and proteins, but with a predominance of protein in the cytoplasm and nucleic acids in the nucleus. This results in different action spectra. This can be demonstrated by numerous examples. If we take as example the division of the sea urchin egg (Fig. 11), the unbroken line shows the action spectrum of the delay in division of the eggs when only the sperm is irradiated. The dotted line shows the effect when only the egg is irradiated.

The spermatozoa are mainly nucleus, and the eggs mainly cytoplasm. In absolute terms the nuclei are much more sensitive, and suffer much more damage. If the nucleus of an irradiated amoeba is replaced by the nucleus of a non-irradiated one, the amoeba becomes normal again, while the amoeba which has received the irradiated nucleus dies.

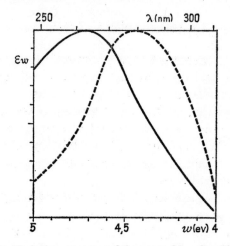

Fig. 11 Action spectrum of the sea urchin, after Giese

The quantum efficiency Q has been studied with particular reference to the bactericidal effect of UV. This is very weak, requiring $3 \cdot 10^6$ photons of 260 nm to kill a bacterium.

The law of reciprocity applies between 10^{-6} seconds and 1 minute. In fact, it is better to define the quantum efficiency by the absorption, not by the whole cell, but by the chromophore responsible for the effect under consideration: in bacterial sterilization, the quantum efficiency then becomes of the order of 1 in 5,000, meaning that 5,000 photons must be absorbed in the molecules where the bactericidal effect is produced, to kill the cell.

The effects of UV radiation on cells depend on the physiological state of the cells; well-nourished cells are much more resistant, thus showing that this is not simply a primary photochemical effect. Furthermore, the existence of temperature-dependent secondary reactions has been demonstrated in the delay in division induced by UV in various ciliate protozoans. The delay involves firstly a retardation before the first division following irradiation; this is independent of temperature and represents the primary photochemical damage due to the photons. There is subsequently a prolonged period known as *stasis*, which begins after one or two divisions have followed

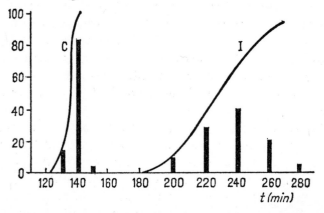

Fig. 12 Loss of synchrony in division of sea urchin's eggs, after Wells

the irradiation. During stasis the cell remains fixed without change. Once stasis is over, the normal rate of division is resumed. Stasis is affected by temperature.

The damage caused by UV irradiation (265 nm) to sea urchin spermatozoa is manifested not only by the delay in division of the eggs already mentioned, but also by the loss of synchrony of cell division which is normally seen in this organism: in Figure 12, curves C and I represent the percentage of eggs which have divided as a function of time t after insemination. C represents the control group of normal eggs, and I represents the eggs fertilized by the irradiated sperm: furthermore, photographs showing details of the degree of synchrony in division were taken every 10 minutes in the case of C, and every 20 minutes in the case of I. Synchrony is very marked in C (more than 80 per cent of the eggs divide at the same time, 140 minutes after insemination), and much less so in I, as is shown by the black rectangles, which give the percentage of eggs, the simultaneous division of which was confirmed photographically.

Occasional reports have appeared of an apparent stimulatory effect of UV, rather than of an inhibiting effect on mitosis (see Gurwitsch 1959). These *mitogenic radiations* are treated with scepticism by most specialists, although in certain conditions a slight increase in the rate of cell division has been confirmed following weak UV irradiation at 280 nm. Even more suspect are reports that cells which are undergoing mitosis themselves give out UV, of very low λ, stimulating division in neighbouring cells.

The *micro-irradiation* of any given region of the cell, by means of a very fine UV beam, is a difficult but interesting technique. It is represented schematically in Figure 13. P (a culture chamber containing cells) is placed on the stage of a microscope whose objective is formed by two mirrors, M and N, the first concave with a hole through the centre, and the second convex. The preparation is seen by visible light, through the eyepiece O and this light enters from below, through the condenser C. In order to irradiate the preparation, light from a UV source, S (for example a quartz mercury vapour lamp), after passing through a filter F, which isolates the selected UV radiation, is concentrated by a quartz lens L through a small hole T, and is then reflected from a glass slide V, the lower face of which is partially aluminized. From the hole T, the objective gives an image in the plane P of the preparation, and the apparatus is adjusted so that this image coincides with the intersection of the cross wires in the eyepiece O. This is possible since the mirror objective is free from chromatic aberrations, so that focusing is the same in UV or in visible light.

44

Using this technique it is possible to verify that the nucleus is the region sensitive to damage which interferes with cell division. The rest of the cell, however, also undergoes alterations.

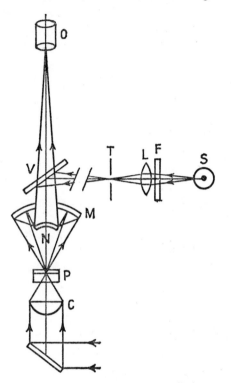

Fig. 13 Apparatus for micro-irradiation

For example the membrane on the cell surface forms a pro-tuberance at the point of irradiation, and may even rupture with loss of cytoplasm and cell death. The pseudopodia of the amoeba retract after irradiation, the cilia of *Infusoria* stop beating and the contractile vesicles of *Paramecia* are immobilized.

The irradiation of chromosomes with a UV microbeam is of particular interest and visible modification of the chromosomes is produced. The action spectrum of this effect shows two

45

maxima at about 225 and 275 nm, with a minimum at 250. There is a diminution in chromosome absorption at 260 nm, which is characteristic of nucleic acids.

The explanation of the more important of the effects just mentioned, in particular the effect on the nucleus of the cell, must be sought in the photochemistry of nucleic acids, at least for the primary process. Nucleic acid photochemistry is conplex and poorly elucidated, although since 1960 considerable advance has been made, particularly in the photochemistry of *deoxyribonucleic acid* (DNA). As DNA plays an essential role in heredity, it would be as well first of all to discuss the mutagenic effects of UV.

3.2. Mutagenic action of radiation

Genetic information is, to a large extent at least, transmitted by the chromosomes of the nucleus, which control enzyme formation in the organism during development. The genetic code is inscribed in the DNA molecules of the chromosomes. DNA molecules are polymers made up from four nucleotides and forming a double helix. Changes known as *mutations* can be observed as chance results of aberrations in the chromosomes, but these are not hereditary. The types of mutation that we are considering are permanent changes that will be transmitted to descendants. In multicellular organisms, these mutations may show themselves anatomically, whereas at a cellular level they are manifested as physiological changes, such as resistance to a particular drug.

In 1927, Muller discovered that ionizing radiation induces mutations in flies, and Altenburg (1933) found that UV had the same effect. These older studies are, however, difficult to interpret. For example, in studying the fruit fly, *Drosophila*, the abdomen of the male fly was trapped between two quartz plates, and irradiated. These flies were then mated with normal females and their descendants studied. However, in these conditions, 99·9 per cent of the UV is absorbed before it reaches the spermatozoa, and if the action spectrum is evaluated in this way, all it really shows is the transparence of the abdomen to UV. Recent work has been mainly carried out on unicellular creatures with a diameter of the order of $0·5\,\mu$ (bacteria, spermatozoa, etc.). The action spectra show a maximum of about 260 nm, and are thus in the neighbourhood of the maximum for DNA absorption. Strictly speaking the position of ε_m should be compared with α_m (which is independent of any convention), in terms of the number of photons and not the energy. This is on condition that the quantum efficiency Q remains constant. Attention is not usually paid to this point, but fortunately the maximum of the curves are sharp and so hardly vary with the chosen convention. Comparison of the curves ε_w and α_λ over a large range of wave-lengths makes little sense if energy values (or rather their inverse) are used for the action spectrum.

For a given monochromatic radiation, the relationship between the *dose* administered (total number of photons during irradiation) and the number of mutations observed is sometimes exponential (and initially even linear, which is the beginning of exponential), indicating that a single photon is enough to produce a mutation, and sometimes sigmoid, which would indicate that several photons are necessary in order to have the same effect. The law of reciprocity is valid (up to 25 minutes' duration in certain cases), and temperature has no effect, suggesting a primary photochemical reaction (this is not strictly so, as we shall see later). The absence of an oxygen effect is generally accepted. It has been shown, however, that in *Staphylococcus aureus* mutations can be produced by culture medium which has been previously irradiated with UV. This is probably due to the formation of organic peroxides, but in this case UV of wavelength below 200 nm is the most effective, and the doses required are very much greater than those more usually employed. We are dealing in this case with a chemical phenomenon different from the direct mutagenic action of UV.

Metabolic processes are known to influence the induction of mutations by UV light. Demerec and Latarjet (1946) showed the existence of delayed mutations which did not appear in the irradiated cells, but in their descendants as far as the thirteenth generation. Witkin (1956) confirmed that mutations could also be produced in the irradiated cells themselves, but with a certain delay; he showed that this phenomenon was related to the presence of amino acids in the culture medium during the first hour following irradiation and before the first division. It is possible that these amino acids react in the irradiated cells with UV modified nucleic acid precursors concerned in nucleic acid synthesis, particularly the synthesis of ribonucleic acid. It is also possible that enzyme reactions, resulting in the repair of DNA molecules damaged by UV, take place in irradiated bacteria.

What is the nature of the damage to the DNA molecule produced by irradiation? The information carried by DNA is based on the sequence of the four nucleotides in the double helix of the molecule. The coding changes if, for instance, permutations are made in the places occupied by two different nucleotides. Thus mutation occurs if a nucleotide occupies the incorrect position, as might happen as a result of a photochemical break in the hydrogen bonds in DNA. In fact the photochemistry of DNA isolated *in vitro* shows that this phenomenon does exist. It is shown by an instability in the helicoidal structure induced by UV so that acids or formaldehyde destroy the molecule more easily than if it had not been irradiated.

Another possibility is a modification of one of the nucleotides. After irradiation of *Enterococcus* at 254 nm, a derivative of thymine, one of the nucleotides, can be detected by chromatography. This substance does not exist in normal cells, and recent work has shown that it is a *dimer* of thymine which can be observed as a reversible photoproduct of DNA *in vitro*. A plane formula of this dimer is shown in Figure 14, another can be

48

obtained by rotation of the right-hand half through 180° around the axis marked by a dotted line. Each of these plane formulae has two corresponding spatial possibilities, according

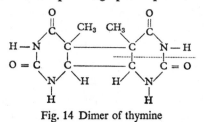

Fig. 14 Dimer of thymine

to whether the two pyrimidine rings are on the same side or one on either side of the cyclobutane grouping formed by the four central carbon atoms of the dimer. Thus there are four possible isomers, and it would be interesting to determine by

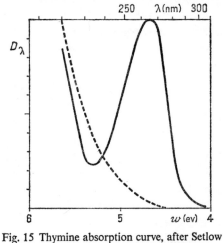

Fig. 15 Thymine absorption curve, after Setlow
———— monomer   ----- dimer

nuclear magnetic resonance which one is formed under a given condition, since this would show whether the thymine radicals which polymerize are in the same DNA chain or in two different ones. The monomer and dimer of thymine have different absorption spectra (Fig. 15), which means the transformation

49

from one to the other can be followed. The proportion seems stable and depends on the radiation used. At 275 nm the dimer is in excess at equilibrium and at 235 nm the monomer is. The quantum efficiency Q is less than 0·01 for the formation of the dimer, and about unity for breakage into the monomer. At 245 nm, at equilibrium there is practically 30 per cent dimer *in vitro*, while in a culture of *E. coli* it is only 10 per cent. This difference may be due to the position of thymine molecules in the DNA helix of the bacterium. It has been reported that equilibrium is dependent on the oxygen concentration in the medium, perhaps because the para-magnetism of oxygen acts on the excitation process, leading to the triplet state.

Although the dimer of thymine certainly plays an important part in the photobiology of cellular DNA, it is not known whether dimerization is the basis of the primary photochemical reaction. The use of chromatography and radioactive carbon labelling of DNA suggests that it is probable that thymine products other than the dimer are formed as a result of irradia-tion, both *in vivo* and *in vitro*. Finally other DNA nucleotides (especially cytosine) are probably affected by irradiation.

The recently studied dimers of pyrimidine appear to play an important part in cellular DNA. A reversible hydration of pyrimidine has also been shown, which could be a mutagenic factor, and also a production of a nucleic acid-protein bridge by a pyrimidine-amino-acid linkage: such compounds have been shown in irradiated viruses, and certain action spectra permit the importance of this reaction to be foreshadowed. It is clear that the problem is very complex and far from being resolved.

3.3. Photodynamic action

This term (or *photosensitization*) has been used to describe the phenomenon whereby a usually inoffensive radiation can damage cells if they have previously been in contact with substances which absorb the radiation. We have already alluded to this important phenomenon. Nearly all photodynamic substances are fluorescent, but the reverse is not true. The presence of oxygen and water is essential, the reaction is irreversible and independent of temperature. It seems therefore to be a true primary photochemical reaction, with a quantum efficiency of about unity. Finally, the order of the reaction is zero with respect to the concentration of the photodynamic substance, while the total number of injured cells is proportional to the number of quanta of radiation absorbed. This proves that the photodynamic molecule acts as a catalyst, passing into the excited state with the absorption of a photon and returning to the initial state after transfer of its excess energy to the cell.

The study of mutations induced by photodynamic action are of particular interest and the first observation in this field was made by Kaplan (1948). He observed that the frequency of mutations in *Bacterium prodigiosum* increased after staining with erythrosine and exposure to light under a microscope, although staining or the influence of light alone were without effect. In experiments of this nature, the action spectrum is obviously related to the absorption of the sensitizing dye.

Mutations, induced by photodynamic sensitization, are biological phenomena of considerable philosophical importance. At the present time light from the sun, reaching the earth's surface, contains only very little energy of less than 300 nm, and the number of photons corresponding to this range is not enough to cause mutations. This would obviously not be the case for a cell in a medium which absorbed visible light. For example sea water shows a maximum transparency of about 480 nm and it also contains yellow pigments, the nature of which are poorly understood. These pigments can absorb solar radiation in this region of the spectrum and induce mutations in marine micro-organisms even at the present time.

Chance sensitizations probably explain some of the unusual

results obtained by various workers who have reported mutations induced by UV between 310 and 400 nm. Light at this
range may produce lethal effects in high doses, but does not
cause mutations.

Although the mechanism is different, photosensitization can
be compared with various chemical sensitizations which aug-
ment the effect of UV radiation on bacteria and even on the
cells of higher animals. Bromouracil, for example, when
incorporated into the DNA of *E. coli*, replacing thymine,
approximately doubles sensitivity to UV, and in the *Enterococcus* sensitivity is increased by about 60 per cent. The action
spectrum shows a flattened maximum between 250 and 280 nm.
There is a considerable increase in the long wavelengths of the
UV spectrum which is doubtless due to differences in the spectral
absorption of bromouracil and thymine. There is only partial
substitution of bromouracil for thymine in DNA (less than a
quarter in bacteria, but this is sufficient to produce an obvious
sensitization, not a physical one as is photosensitization, but biological: the enzyme which splits off a thymine dimer cannot recognize
and therefore cannot split off a thymine bromouracil dimer.

Photosensitization is a physical and not a biological process,
therefore it applies also to macromolecules which, like viruses
and enzymes, are on the borderline of life. Irradiation causes
photo-inactivation, which has physiological repercussions. The
tobacco mosaic virus has been much studied. The action spectrum is unrelated to the absorption spectrum, since the quantum
efficiency varies with λ (in addition it is essential to take into
account diffusion, which is very important and vitiates absorption measurements). It is also possible that the abnormal action
spectrum is due to the inactivating lesions being mainly due to
RNA-protein bridges of the type mentioned at the end of **3.2**.
Stains such as acridine orange or acriflavine can inactivate
viruses under the influence of visible radiation; here again the
importance of this phenomenon is stressed from both theoretical
and practical points of view.

The details of the mechanism of photosensitization are not fully known and seem also to be quite variable. Dyes of the fluorescein type stay on the surface and act on the proteins in the cytoplasm, especially on their aromatic amino acids. Dyes of the acridine type, on the other hand, combine with nucleic acids, and can also be mutagenic. Finally, naturally occurring photosensitizers have been found in certain cells, for example the red ciliate protozoan *Blepharisma*, and in some purple bacteria.

3.4. Photoreactivation

In 1948, Kelner discovered that the lethal effects of UV on the cell could be repaired by subsequent or simultaneous irradiation with visible light. This phenomenon of *photoreactivation* or *photorestoration* can easily be confirmed by making use of the capacity of most micro-organisms to divide and multiply into a colony. Reactivation is not complete and it reaches an asymptote when the duration or the intensity of the restoring postillumination is increased.

Figure 16 illustrates this and shows the surviving fraction in a culture of *Escherichia coli* (that is, the proportion of cells which retain the power to reproduce by division) as a function of the dose of UV irradiation; both scales are logarithmic. The curve N refers to the usual results, obtained in the dark, and the curve P to the effect of a photoreactivation due to exposure for 1 minute to an incandescent lamp. The curves can be superimposed by moving them along parallel to the axis of the abscissa. This indicates that the surviving fraction for P can be obtained by multiplying the dose for N by a *dose reduction factor* (DRF) which in this example is about 2. This result is not generally applicable, but is a convenient approximation in many cases. The factor depends on the organism and the lethal effect of the UV under consideration.

The action spectrum of the photoreactivation depends on the culture under experimentation. In *Streptomyces griseus*, for example, which is the organism in which Kelner discovered the phenomenon, the action spectrum shows a sharp maximum at about 435 nm. The action spectrum of *E. coli* is clearly different (Fig. 17); it starts at 310 nm, which is exactly the radiation at

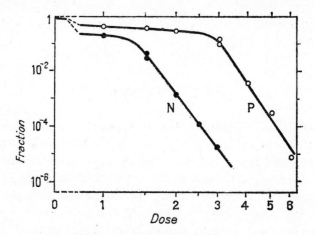

Fig. 16 Photorestoration, after Kelner, showing
surviving fraction and dose

which the lethal action spectrum of UV ends in this case, and extends into blue visible light (500 nm), with maxima at about 350 nm. It is therefore the UV close to the visible, rather than visible light itself, which acts as a photorestorer. One difficulty in the determination of action spectra is that the law of reciprocity is usually inapplicable. At low intensities of reactivating light, acting over a long period, it appears that the possibility of reactivation falls progressively with time after UV irradiation (in several hours it becomes negligible, except where metabolism is reduced to a minimum by cold and a non-nutritive medium).

These departures from reciprocity suggest that photoreactivation involves not only the initial photochemical process, but ill-understood secondary reactions. This is suggested by the fact that temperature acts on photoreactivation. Oxygen is apparently not involved.

The mutagenic effect of radiation is also susceptible of photo-reactivation, but in a slightly different way. For example, in *E. coli*, photorestoration of the capacity to divide falls expon-entially to zero in about 3 hours at a temperature of 20° C,

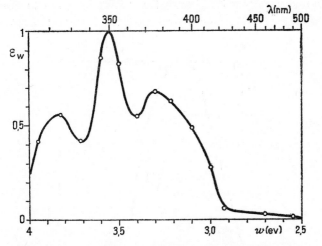

Fig. 17 Photorestoration of *E. coli*, after Jagger and Latarjet

whereas for the same organism restoration of the mutagenic effect is abolished in 20 minutes.

Many other phenomena induced by UV in micro-organisms are also susceptible of photoreactivation: for instance the slowing down of division in paramecia and the immobilization by UV irradiation of some unicellular organisms, where a reactivating action by UV of wavelength less than 240 nm, has been described following immobilization by 260 nm.

It is not impossible that this paradoxical phenomenon is less exceptional than it seems, since usually UV of shorter wave-length than that for irradiation also has lethal effects which mask a possible restoring effect. Another curiosity is the *photoprotection* which is occasionally observed and in which the reactivating radiation is applied before that which causes the damage. The action spectrum for the lethal effects on *E. coli* (Fig. 18) differs altogether from that of normal photoreactivation (Fig. 17). In addition, the law of reciprocity applies and there is no temperature effect. The mechanism is therefore different.

In the case of viruses, photoreactivation has been particularly studied in *bacteriophages*. The 'plaque', that is, the region of

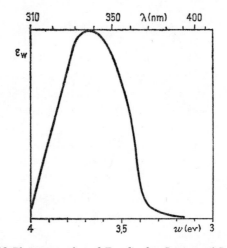

Fig. 18 Photoprotection of *E. coli*, after Jagger and Stafford

destruction in a colony of bacteria on a plane culture medium, is very sensitive to UV and irradiation can be carried out before or after infection without appreciable difference. On the contrary photoreactivation is only effective after infection of the host. In the case of *E. coli* infected by a bacteriophage, the action spectrum of photoreactivation is the same as for division of the hosts themselves (see Fig. 17, p. 55), as is the temperature effect. The identical nature of these mechanisms is important in considering the theoretical interpretation of photoreactivation. It is in fact known that infection is due to the injection of viral DNA into the host. Inactivation of the phage by UV is due simply to the damage caused to the DNA by UV, which deprives it of effect on the host. The bacteriophage, with impaired metabolism, cannot repair this damage, whereas once the damaged DNA is incorporated into the bacterium, it can utilize its photoreactivating mechanism to restore the DNA to its non-irradiated form (which is not in its own interest, since it will die from it!).

An initial photochemical mechanism of rapid growth, proportional to the illumination and little affected by temperature, can be identified by the detailed study of the kinetics of reactivation of the phage as a function of the duration of illumination. Following this is a slower exponential effect, due to the fact that a substance essential for the first reaction becomes exhausted and the cell then has to produce it by an ill-understood temperature-dependent reaction. In strong lights, the second process limits the possibility of restoration to an asymptomatic value, whatever the intensity of the photo-reactivating light. Because of this, light applied as a series of flashes separated by periods of darkness is more active than a single long exposure of equal composition and energy. If the temperature is lowered from 40° C to 0° C just after a flash, restoration is not affected, but, if it is lowered beforehand, there is a decrease in photo-reactivation. The longer the duration of the temperature drop the more marked is the decrease and this is true, up to a limit of about 30 minutes, which is the time interval necessary at 0° C, for the action of successive flashes to become independent.

These complex phenomena are interpreted in the following way. UV causes two different alterations in DNA, since photo-reactivation is never complete. The irreversible damage corresponds to about a tenth of the applied UV dose, the reversible damage is repaired by an enzyme according to the following scheme:

$$E + A \rightleftharpoons EA \xrightarrow{h\nu} E + R$$

where E represents the enzyme, which becomes reversibly

attached to the altered molecule A of DNA to form a complex EA, which is destroyed by light, liberating the enzyme and the repaired molecule R of DNA. The enzyme E itself is produced in the cell from a certain substance with which it is in equilibrium (dark, temperature-sensitive reaction). This sequence of events has been demonstrated by extracting an enzyme from yeast and using it *in vitro* to facilitate the light restoration of DNA previously damaged by UV irradiation.

The theory that thymine dimer is involved in the DNA lesion which can be photoreactivated (as already discussed above) is at present in favour. If DNA is irradiated with UV and subjected to hot-acid hydrolysis, the dimer can be demonstrated in the mixture, whereas if the DNA has been previously illuminated, after it has been in contact with partially purified preparations of yeast enzyme, much lower, or even negligible, concentrations of dimer are obtained. As we have already mentioned the partial substitution of bromouracil as a replacement for thymine strongly increases the sensitivity to UV of organisms or of free DNA. In these two cases photoreactivation is very clearly diminished, which confirms the part played by thymine in the lesion which can be repaired by light.

Other less direct arguments can be used to support the dimer hypothesis. The absorption of a photon at 254 nm per nucleotide into DNA would damage less than 1 per cent of the nucleotides; the quantum efficiency Q is thus very small and is practically of the same order of size as the concentration of the dimer of thymine recovered from the DNA hydrolysates.

The dimerization of thymine under the action of 280 nm is destroyed at 239 nm which splits the dimer into monomers. The paradoxical photoreactivation which results is due to ultraviolet of shorter wavelength than that which caused the damage and this can be explained by a purely photochemical mechanism. This has in fact been confirmed, as already described, in experiments on DNA with or without treatment with the photorestoring enzyme.

Although many of the details of photoreactivation are still obscure, the mechanism of this important phenomenon has begun to be unravelled. It should, however, be realized that experiments on artificial DNA systems *in vitro* may differ considerably from the effects produced in living cells. In the laboratory, high doses in the order of tens or hundreds of photons per DNA nucleotide are often employed whereas biological damage requires only one photon per several nucleotides. In the first case the probability of effects where several photons play a part in the same reaction is therefore appreciable. This is not so in the second case. In spite of this there are numerous findings which point to a close resemblance between natural photoreactivation and that observed experimentally on isolated DNA. For example, a non-photorestorable mutant of *E. coli* has been found, whose extracts lack the enzyme needed to repair artificial damage to DNA, whereas the original line, used as control, possesses the enzyme.

It is, however, far from proven that all UV damage acts through DNA and it is certain that other mechanisms exist. For example one strain of *E. coli* reacts differently to photosensitization by acriflavine, followed by photoreactivation, depending on whether the effects are considered as lethal or mutagenic. Very little is known about photoreactivation in which DNA is not involved, and nothing at all about photoprotection.

The evolutionary significance of photoreactivation is an important question. In spite of the low concentration of harmful or mutagenic UV in solar radiation at the present time, it is capable of exerting a definite effect on DNA. If a culture of *H. influenzae* in a quartz container is exposed to the sun in summer for one hour, its resistance to streptomycin decreases about 100 times. If the photoreactivating yeast enzyme is then added, considerable regaining of resistance is observed after the same exposure to the sun. Even at the present time photoreactivation could thus be useful in protecting some species, in particular against mutations. At the beginning of life on our planet, where solar UV of short wavelength must have had greater energy, it is possible that various photoreactivation mechanisms were operating to protect living creatures from both its lethal effects and the widescale induction of mutations. That which is observed today is only the reminder of a time when photoreactivation was indispensable to life. This was originally created under the influence of UV and would have risked being destroyed by it were it not for the beneficial interaction between UV and visible light.

Finally it is worth noting that in some bacteria photoreactivation does not occur. These cases, fairly scarce, seem to occur particularly in bacteria whose DNA undergoes *in vitro* transformation. This coincidence has recently been shown in many of the cells of higher organisms which never see the light. Professor Latarjet thinks that the enzymes of photoreactivation act to correct the errors which appear spontaneously, due to thermal agitation, after replication of DNA. Photoreactivation would thus be a particular example of a general biological phenomenon, stabilization of the genetic inheritance by enzymes with the ability to correct certain accidental mistakes.

REFERENCES

DEMEREC, M., and LATARJET, R. Cold Spring Harbor Symposia Quant. Biol., **11**, 38 (1946)

GIESE, A. C., Biol. Bull. **91**, 81 (1946)

GURWITSCH, A. and L., *Die Mitogenetische Strahlung, Iena, Fischer* (1959)

HOLLAENDER, A., ed., *Radiation Biology*, vols. II and III, N.Y. and London, McGraw-Hill, 1955 and 1956

JAGGER, J. and LATARJET, R., *Ann. Inst. Pasteur*, **91**, 858 (1956)

JAGGER, J., and STAFFORD, R. S. Photochem. and Photobiol., **1**, 245 (1962)

KAPLAN, R. W., Naturwiss. **37**, 127 (1948)

KELNER, A., J. Bacteriol. **58**, 511 (1949)

LEA, D. E., *Action of Radiations on Living Cells*, 2nd ed., Cambridge University Press, 1955

SETLOW, R., Biochem. et Biophys. Acta, **49**, 237 (1961)

WELLS, P. H., Ph.D. Thesis, Stanford University (1952)

WITKIN, E. M., Proc. Nat. Acad. Sci. U.S., **32**, 59 (1956)

CHAPTER IV

Photoreception in Animals

4.1. Phototaxis

In considering the increasing complexity of the animal kingdom passing from unicellular organisms to, on the one hand, socially organized insects and, on the other, to the primates and man it is convenient, having looked at the effects of UV light on cells, to examine the response of simple micro-organisms to light.

One of the most striking features is the tendency which living creatures possess of moving either towards or away from a light source.

Phototropism in plants (de Candolle, 1832) is a mechanical action resulting from the unequal distribution of *auxin* (growth hormone) in the stalk. The concentration is higher on the side away from light due to transport laterally, the nature of which is not fully understood.

Movements of insects and vertebrates towards or away from light have long been familiar (Loeb, 1888). These are complex behaviour reactions connected with a visual receptor system. We shall return to this point later on. By contrast it is only recently that more rudimentary reactions have been studied among unicellular or simple multicellular organisms. These responses have been grouped under the term *phototaxis*. Photo-taxis is exhibited mainly in organisms which are classified as plants (photosynthetic bacteria or algae) but, at this level of organization, the distinction is fine.

The earliest documented case is that of purple bacteria (Engelmann, 1833) where phototaxis is described as *phobic*, indicating that the motor response is controlled by change of the light intensity in time without influence from the direction of incidence. For example *Rhodospirillum* swims by means of flagellae (which act rather like propellers) until the light intensity is decreased; the bacterium stops swimming for one or two seconds, and during this rest its orientation changes due to Brownian movement, so that it sets off again in a slightly different direction. The result is that by trial and error the bacteria accumulate near the light source. It thus appears as though the micro-organism has swum in the direction of the source, although in reality the direction only acts through the light gradient at various distances from the source, and not by true orientation in the direction of the light rays. This can be demonstrated by concentrating the light from a source through a lens; the bacteria accumulate where the light intensity is greatest and may travel in the opposite direction from the light source to do so.

Phototaxis in these bacteria is an example of the general biological phenomenon of irritability and is similar in many ways to the response of nerve cells to a stimulus. In this, which we will examine in the case of vision, the time scale differs from that seen in bacterial phototaxis. In the latter case the time scale is measured in seconds whereas in the nerve a comparable event is measured in milliseconds. We are, apart from these exceptions, dealing with general laws of nervous excitation: refractory period, recovery, rhythmicity, adaptation, Weber's law (proportionality between the just effective variation in light intensity and the intensity itself), Gildemeister's effect (a stimulus too weak to be effective which can nevertheless inhibit a subsequent stronger stimulus).

Phototaxis in purple bacteria changes its sign in very strong light intensities and the bacteria move away from the light source. The same effect may be observed with colourless bacteria which have been stained with a dye, through a photosensitization phenomenon.

The purple bacteria which exhibit phototaxis are equally

capable of photosynthesis, due to chlorophyll and carotenoids which they contain. The action spectra of these phenomena are identical. The phototactic response of the bacteria is simply the result of a sudden decrease in photosynthesis, and a reduction in the quantity of *adenosine triphosphate* (ATP) supplied to the flagellate mechanism. This shock reaction affects the movements.

Purple bacteria do not possess a localized photoreceptor, and therefore constitute the most rudimentary form of response to light. They are usually found under a layer of algae which only allows the near infrared (750–950 nm) to filter through, and it is precisely in this region of the spectrum that the chlorophyll of the bacteria is absorbent. A single pigment is sufficient to maintain their metabolism and movement. They exhibit *phobo-phototaxis* when they are left in contact with a layer of algae in the position where they will receive the maximum amount of light. If, however, the light becomes too strong and there is a risk of damage to the bacteria, they retreat and the effect is reversed. Even the opponents of finality cannot fail to admire the surprising economy of means for achieving a result so well adapted to the environment.

Our second example is taken from a green alga, *Euglena*, whose characteristics are quite different from those of purple bacteria. This organism exhibits *topo-phototaxis*, which means that it moves towards the source of light. Here the direction of the light rays reaching the cell is the most important factor determining movement of the organism and not the light gradient as in the preceding example. *Euglena* moves towards a beam of light rays converging through a lens although the light diminishes as it progresses in that direction. To understand the mechanism whereby it finds the direction of the source, it is necessary to consider the cytology of *Euglena*. The cell swims by means of a flagellum situated towards the front in the direction of progression. Near the base of this flagellum is an orange spot pigmented by a carotenoid, known as the *stigma* (sometimes, quite wrongly, referred to as the 'eye'). The stigma is not the organ of photoreception. This is situated in a thickening at the base of the flagellum and no absorbent pigment is seen in the region, probably because the concentration is too weak. It is probable that the directional sense is due to the shadow thrown by the stigma over the photoreceptor organ when these organs are in line with the source of light. This arrangement is much more elaborate than that of purple bacteria, and it is concentrated at two points in the cell instead of acting throughout its volume. Here also aversion to very strong light intensities occurs and the organism retreats from the light.

The action spectrum also differs from that of the preceding example. Chlorophyll plays no part, and it is only blue visible light that is active, and not infrared. The best way of determining the action spectra is to place the organisms in a container lit in two opposing directions, a monochromatic beam in one direction being balanced by a constant comparable beam in the other. The energy of the first beam is varied until equilibrium (the absence of systematic movement) is reached. This action spectrum shows a main maximum at 495 nm and a secondary maximum at 425 nm. It is possible that selective absorption by the stigma, which varies from 40 per cent for short wavelengths in the visible range to 10 per cent for long wavelengths, influences the results. In strong light intensities the action spectrum seems to remain the same for the negative phototaxis which is then observed, unless the organisms have been previously adapted to darkness for several hours, in which case the action spectrum shows a single maximum at 140 nm. It is possible that in this last instance the stigma no longer functions and that a phenomenon of phobo-phototaxis is seen, the organism being 'dazzled' and moving away from the light because of gradient and not orientation. Measurements of the action spectrum on a mutant of *Euglena* not possessing a stigma support this explanation. The positive reaction was weak with a maximum of 410 nm, the likely absorption maximum of the photoreceptor pigment.

Negative phototaxis of *Euglena* has been recently studied by Diehn and Tollin (1969) using polarized light: the shading effect of the stigma changes with the orientation of polarization, because the pigment molecules in this organ are aligned with their long axes parallel to the long axis of the organism. The action spectrum so recorded shows major peaks at 450 and 475 nm and a minor one at about 415 nm, and it seems that it may be explained by the existence in the stigma of a carotenoid shading pigment. In the photoreceptor the pigment is perhaps a flavonoid.

Although the reversal of phototaxis is linked essentially to the light intensity, it appears that light modifies the intracellular distribution of cations and that reversal is mediated by changes in the concentration of K^+, Mg^{++} and Ca^{++} ions within the cell. Some algae, for example, show positive phototaxis in solutions where the ratio of the numbers of magnesium and calcium ions is greater than 6, and negative phototaxis where the ratio is smaller. The activity of ATP in muscular contraction in higher animals is related to the concentrations of cations, and this again suggests that, as in bacteria, ATP is an intermediary between the action of light and movement. It is difficult to describe the mechanism of these actions in more detail, since the nature of the light-sensitive pigment in *Euglena* and similar micro-organisms is still unknown. It is possible to blind *Euglena* and abolish phototaxis by micro-irradiation with UV of 280 nm concentrated on the photosensitive region, showing that the receptor pigment has been destroyed.

The study of *Euglena* is complicated by the fact that in addition to phototaxis it also shows *photokinesis*, that is the speed of swimming is influenced by the level of light intensity, without in this case reference to the direction of the light. The action spectrum of photokinesis differs from that of phototaxis; it has a principal maximum at 465 nm and a flattened secondary maximum at about 600 nm which is related to the absorption spectrum of chloroplasts in *Euglena*. In this organism chlorophyll operates in photokinesis, but not in phototaxis

65

(Fig. 19). Photokinesis provides a simple method of estimating variations in ATP synthesis. *Euglena* already, therefore, seems to possess a differentiated group of mechanisms which allow it to approach light sources by phototaxis, the speed of approach increasing as the distance diminishes, due to photokinesis. These two mechanisms, however, are not entirely independent, despite their different origins. The threshold of phototaxis is

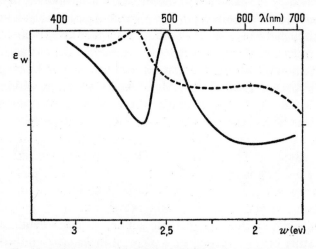

Fig. 19 Action spectra of phototaxis (solid line) and photokinesis (dotted line) of *Euglena gracilis*, after Wolken

raised by light adaptation and this effect is due to a decrease in in the cellular CO_2 tension as a result of photosynthesis.

There are, of course, other forms of phototaxis than those which have been illustrated by the two examples of purple bacteria and *Euglena*. For instance, diatoms move on rock surfaces by means of a directional mucus excretion. The time scale is slow and it takes several minutes before response to light occurs, whereas in the preceding examples it was a matter of seconds. It appears that both photo- and topo-taxis operate, but the action spectrum has not been worked out.

4.2. Photoreceptors in invertebrates

If we pass on from unicellular organisms to the multicellular *Metazoa*, photoreception becomes extremely diversified. It is not within the scope of this book even to summarize the fascinating story of the evolution of photoreceptor organs in animals, and we can only cite a few examples to illustrate the main points.

Almost all lower animals possess, at the surface of the body, cells with a special sensitivity to light. These cells are sometimes spread over the whole surface and sometimes concentrated in various regions. In the earthworm there are concentrations towards front and back. The animal 'sees' (if it can be called seeing) from its two extremities. The photoreceptor cells are recognized by the presence within them of a *black pigment* which absorbs light. The pigment is concentrated to the side opposite to that through which light penetrates. The function of this more or less generalized pigmentation is as yet unexplained and suggestions as to its photoreceptor role in the animal have not been substantiated.

In the next stage cells are grouped to form *ocelli* which constitute the most rudimentary eyes. The flat ocelli of worms are the simplest form. In molluscs and insects the ocelli are hollowed out into a cup shape and frequently the epidermis closes the cup towards the front by a transparent thickening and the rough lens so formed causes light to converge. This lens does not form an image, but it does, to some extent, concentrate the light rays. It is more effective than the flat ocellus which can only recognize the direction of light differentially (by comparison of the responses to two different orientations). The cup-shaped ocellus, on the other hand, since it possesses a light condenser, is an effective organ for topo-phototaxis in the animal.

In some invertebrates, however, such as cephalopods, there is a transformation into an image-projecting organ which is very similar to the vertebrate eye, although embryonic formation is quite different. The close resemblance between the eye of an octopus and the mammalian eye is one of the curiosities of nature.

Arthropods have compound, or faceted, eyes. These convex eyes are made up of juxtaposed *ommatidia* in numbers which vary according to the species, ranging from a few units in certain ants to more than 20,000 in the dragonfly. It is usual to distinguish two sorts of compound eye, according to whether the

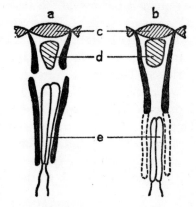

Fig. 20 Apposition eyes (a), and superposition eyes (b), after Exner

ommatidia are of the type described as *apposition* or as *superposition* (Fig. 20). In both cases light is first concentrated by a *cornea c* and then by a crystalline cone *d*, on to the *rhabdom e* which is formed of a group of from 3 to 12 contiguous photoreceptor cells. The organization of the sheath of absorbent pigment surrounding the ommatidium is such that in *a* (apposition type) the only light to reach the rhabdom has come through the cornea of the same ommatidium, whereas in *b* (superposition type) the rhabdom can receive light from surrounding ommatidia. These structures are characteristic of diurnal and nocturnal insects respectively. Moreover, when the latter need to see in light, a migration of pigment (dotted lines, Fig. 20b) isolates the rhabdoms of superposition eyes and changes them into eyes of the apposition type. These explanations are hypothetical and it seems likely that the migration of pigment is a protective process against excessive light. This is a frequent phenomenon throughout the animal kingdom, and is sometimes replaced (particularly among mammals) by a contraction of the pupil of the eye.

Little is known about the dioptric of compound eyes. Since the work of Exner (1891) the lens had been thought of as a heterogeneous structure with major variations of refractive index between its centre and its periphery. Measurements made with an interference microscope show that this is not the case and the lens is homogeneous. In the shrimp, *Aristeus antennatus*, the refractive index of the lens is 1·390 at 585 nm whereas that of the cornea is 1·547 (Carricaburu, 1966). Whereas the vertebrate crystalline lens is clearly an image-forming device, it seems most probable that in an ommatidium the lens concentrates light in directions of its axis by internal reflections. In many day-flying butterflies there exists a thin cylindrical extension of the conical lens, the *crystalline tract*; its diameter varies according to the species between 2 and 10 μ and its length between 20 μ and 1 mm. As the refractive index of the tract is higher than that of the surrounding medium, the tract may function as a *light guide*. Another point of interest in the compound eyes relates to the propagation of rectilinearly polarized light in individual ommatidia. The behaviour of many animals with compound eyes indicates that they are sensitive to the polarization of light, whereas in vertebrates this sensitivity is virtually non-existent. Figure 21 shows birefringence in the cockroach eye corneule. The collected corneules (*cuticle*) form a regular pavement which is in this case rectangular but is sometimes hexagonal. If the tissue is stripped of material which normally lies behind it (lens, rhabdom and interstitial medium) and examined on the stage of a polarizing microscope the cross characteristic of birefringence in convergent light is seen in each corneule. It is possible that this birefringence is a factor in the vision of polarized light, although it is not known how. The analysis of polarized light has been attributed to the structure of the rhabdoms.

A rhabdom consists of a collection of several visual cells, or *rhabdomeres*, which according to the species are 1 to 2 μ in diameter and 50 to 100 μ in length. Each rhabdom appears to

50 μ

Fig. 21 Birefringence of corneules in *Periplaneta* (Physics Laboratory of the French Natural History Museum)

be neurally independent since each retinular cell gives rise to its own nerve fibre. The rhabdomeres may be contiguous ('closed' type), or an axial cavity ('open' type) filled with a transparent liquid may be left between them. A fine structure can be identified in each rhabdomere under the electron microscope. This is characterized by small tubes which are closely approximated. A section perpendicular to the axis of the ommatidium, that is, along the length of the rhabdomere, shows only a lamellar structure (Fig. 22, section AA') whereas in an oblique cut, sections of tubes are seen in some rhabdomeres (section BB'). Each tube is about 50 nm in diameter, with a lining of double membrane from 5 to 10 nm thick. There may be up to 100,000 tubes in a single rhabdomere. Figure 22 represents diagrammatically a rhabdom of the 'closed' type

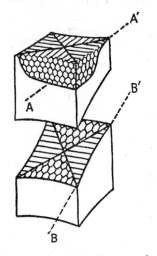

Fig. 22 Fine structure of rhabdomeres, after Wolken

composed of four rhabdomeres. The type of symmetry which is seen in rhabdomeres may perhaps make them more likely to be differentially stimulated by rectilinearly polarized light after orientation of the electric vector. The electric, not the magnetic field, is in fact the basis for photochemical, photographic and fluorescent effects. This has been established by experiments using stationary waves, where the nodes of the magnetic vector have been found to correspond to the antinodes of the electric vector. It would seem reasonable to extrapolate this result to photobiology in general. This theory of the sensitivity to polarized light is challenged by a new one which is based upon

71

the 'modes' of wave propagation in a light guide such as the crystalline tract. We shall return to this point later.

4.3. Photoreceptors in vertebrates

Unlike the eye in invertebrates, which is a specialized part of the skin, the eye of vertebrates is an expansion of the brain. In fact

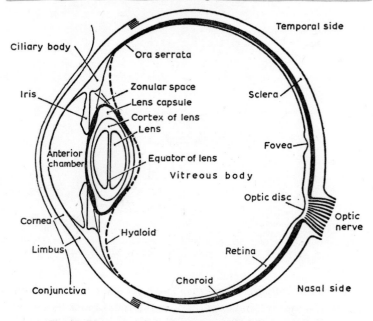

Fig. 23 Diagrammatic horizontal section of the human eye

it is a small peripheral brain. This explains complex innervation of the retina and also its curious inverse arrangement whereby the photosensitive cells are found in the deepest layer of the retina.

The construction of the vertebrate eye is a variant of one model which shows surprisingly little variation. Diagrammatic sections of the human eye (Fig. 23) and the retina (Fig. 24) are included to freshen the reader's memory. The photoreceptor cells, *cones* and *rods*, do receive light only after it has passed through relay nerve cells. Relay cells comprise *bipolar cells*

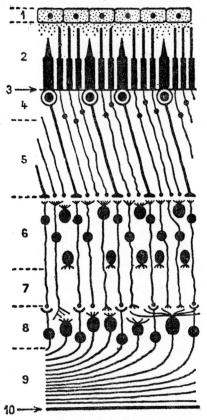

Fig. 24 Diagrammatic section of the retina (light arriving from below)
The following are important layers:

1. pigmented epithelium	6. bipolar cells
2. cones and rods	8. ganglion cells

(which generally receive information simultaneously from a few cones and many rods) and *ganglion cells* (each one is connected to several bipolars). The axons of the ganglion cells make up the fibres of the optic nerve which transmits the visual impulse to the brain. Clearly this arrangement does not favour the quality of the retinal image which is projected by the cornea and lens, since there is a risk of losing detail due to light scattering in these tissues. Therefore, in the *fovea*, a small depression in the retina which corresponds to high-accuracy central vision, not only are the cones (the only photosensitive cells in this region) very closely packed together, but the relay cells are piled up around the fovea, forming a rim.

Image formation in the vertebrate eye is, unlike the invertebrate eye, fairly well understood and we will, therefore, confine ourselves to the question of transparency as this is important in connection with visual pigments. By some ill-explained mechanism the normal cornea is remarkably transparent throughout the visible spectrum. The aqueous humour, which is salt water, is also absolutely clear. The vitreous body, on the contrary, diffuses light considerably, but does not usually show a markedly selective absorption. The lens, on the other hand, exhibits a considerable variation in its transmission factor τ_λ which shows a marked fall in the short wavelengths (Fig. 25) and disappears altogether in UV. As a result of this the lens constitutes a natural barrier which prevents UV from reaching the retina and thus from being active on visual pigments. The physiological advantage of this suppressible process lies in the *chromatic aberration* of the eye, which cannot be corrected because all the transparent media of the eye have practically the same dispersion, that of salt water. Because of this, if the eye is in focus in the middle of the visible spectrum (yellow-green) it is already distinctly myopic at the violet end of the spectrum, and would be even more so in UV. In the compound eye of an insect, where, properly speaking, no image is formed, this inconvenience disappears. Insects are in fact sensitive to UV, at least that reaching the earth's surface from the sun.

Another, and somewhat disadvantageous feature of the lens is the fact that it is constantly modified with age. It is the only organ which starts ageing at birth, since it does not possess a blood supply. The hyaloid artery which supplies the foetal lens is obliterated and resorbed two months before birth. The lens thus progressively loses its plasticity, whence the decrease in the power of accommodation (*presbyopia*). It also gradually becomes yellow, so that in old age violet and blue rays are sparingly transmitted (see Fig. 25, below). The substance which causes this decrease in transparency to short wavelengths, and

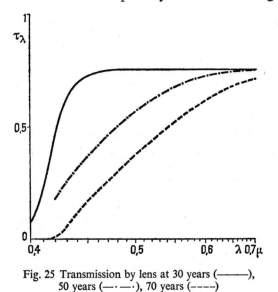

Fig. 25 Transmission by lens at 30 years (————),
50 years (—· —·), 70 years (----)

which has a darkening effect on blues among old people, is a protein which has recently been isolated.

Another yellow pigment, the macular pigment, is present in the retina. This surrounds the foveal region, and probably also serves to diminish chromatic aberration. It does not vary with age, but its concentration differs considerably from one subject to another in relation to the general pigmentation. The black races have it in large amounts, whereas a blue-eyed blond from a white race would have hardly any. This pigment, which is known as *xanthophyll*, is the same pigment which turns the

75

leaves yellow on the trees in autumn. At its absorption maximum (about 450 nm) the xanthophyll has an average optical density D_λ of the order of 0·5. This is by no means negligible and it interferes with the spectral photosensitivity of the retinal receptors in a way which varies from one observer to another. It is possible to observe macular pigment in oneself by means of the entoptic phenomenon known as *Maxwell's spot*. If, on awakening, when the retina is thoroughly rested, a uniform surface such as the bedroom ceiling, is gazed at, a darker patch is seen for a few seconds. This patch is elliptical, with the major horizontal axis about the point of fixation, and it corresponds to the macular pigment. The phenomenon is more obvious if seen through light blue glass. It then disappears, due to adaptation, like all the permanent elements in the visual field, for example the shadows of the retinal vessels.

Oily coloured drops also exist in the retina of some animals, but since they are sometimes thought to play a part in colour vision, they will be considered later.

Polarized light does not appear to have any definite effect on vertebrates. In man, however, Haidinger (1844) described the 'brushes' which are named after him and which are seen in rectilinearly polarized light. They have diabolo-like appearance, yellowish surrounded by blue, and turning with the plane of polarization. They were thought of for a long time simply as curiosities but they now have a medical application in the treatment of strabismus, since they show the normal fixation point (at the centre of the fovea) and allow it to be re-established by exercise in some subjects with abnormal fixation. The classical explanation involves dichroism, that is, an absorption which depends on the state of light polarization in the fibres ending the fovea. In circularly polarized light, some subjects still see tufts in certain directions; this is probably due to corneal birefringence which may transform circular polarization into a rectilinear one.

We will now go on to consider briefly the structure of photosensitive cells in the retina of vertebrates. Since Schultze (1866) it has been known that two types exist, *cones* and *rods*, so called from the shape of the outer segment which is in contact with the pigmented epithelium. Schultze found that the rods must possess a laminar structure, since they disintegrate easily to form stacks of small discs. Schmidt's studies (1938) on the dichroism of rods show that the molecules responsible for light absorption must be in parallel orientation in the small discs. Work carried out

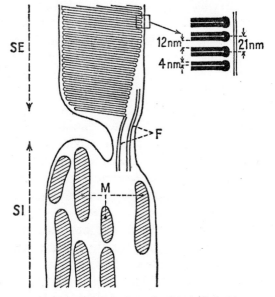

Fig. 26 Diagram of a rod, after de Robertis

since 1949 using the electron microscope has shown that vertebrate rods are formed by superimposed discs about 20 nm thick alternating with less dense layers of 20 to 50 nm thick. Each disc appears to be composed of a double lamella whose membrane is 5 to 8 nm thick. The discs constitute the external segment SE, while the internal segment SI contains the *mitochondria* M which produce the enzymes necessary for metabolism of the rod (Fig. 26). The two segments are separated by a continuous membrane, crossed by fibrils F which link the

77

segments. These almost certainly play some part in the chemical functioning of the rod, since they join the enzyme production centre to the photochemical storage depot which forms the outer segment, as we shall see in the next chapter. It is interesting to note that, from the embryological point of view, rods and cones are derived from flagellate formations similar to the fibrils found in ciliates, the tails of spermatoza and numerous other animal cells. In addition to this ciliary connection between the inner and outer segments, there seems also to exist, at least in some mammals, a cytoplasmic bridge through which the cytoplasm of the inner segment comes into direct contact with the discs of the outer segment.

The structure of the cones is less well known than that of the rods; they are in addition more fragile. Their outer segment also seems to be made up of superimposed discs, which, instead of all having more or less the same diameter, diminish with increasing distance from the inner segment. This accounts for the conical rather than cylindrical form of the outer segment. Also, in the retina of any particular animal species, the cones show a much greater variation in form than the rods. In man, for example, the rods are uniformly 60 μ long, and 28 μ in the outer segment. The diameter varies between 1 μ for the most central rods (those nearest to the fovea) and 2·5 μ for the most peripheral. The foveal cones are thin like rods (75 μ long, 40 μ in the outer segment, with a maximum diameter of 2 μ) whereas those on the extreme periphery are short and fat (21 μ long, 6 μ in the outer segment, and 9 μ in diameter). The variation is continuous from the fovea to the periphery.

REFERENCES

CARRICABURU, P., Vision Res. 6, 597 (1966)
DIEHN, B., Nature, 221, 366 (1969)
EXNER, S., Die Physiologie der facettieren Augen von Krebsen und Insekten, Vienna, Franz Deuticke (1891)
MILLER, W. H., BERNARD, G. D. and ALLEN, J. L. Science, 162, 760 (1968)
DE ROBERTIS, E., J. Gen. Physiol. 43, Suppl. 1–13 (1960)
SCHMIDT, W. J., Kolloidzschr. 85, 137 (1938)
WOLKEN, J. S., Vision, Springfield, Thomas (1966)

CHAPTER V

Visual Pigments

5.1 Experimental techniques

The first visual pigment was discovered by Boll (1876) in the frog's retina. He detached it from the epithelium and found that observing light through it the initially pink coloration became rapidly yellow. The pigment, which is unstable to light, is a coloured substance (i.e. it has an absorption spectrum in the visible range) and the colour disappears in light due to a chemical change which modifies the absorption spectrum. Boll called this pigment *visual purple;* it is now more commonly known as *rhodopsin.*

This discovery aroused considerable interest and stimulated much research. Visual purple is only present in the rods. It exists in the human retina, its presence being first confirmed in the retina of a criminal executed in the dark of 1877. Kühne showed in 1878 that a solution of biliary salts would liberate the purple from the rods of retina macerated in the solution, so that it became possible to study rhodopsin *in vitro.* He also discovered in the fish retina a second visual pigment, *porphyropsin.* At the present time about a hundred visual pigments are known.

There are three different methods of studying visual pigments. The first, which is the oldest and still the most frequently used, consists of extracting the pigment from the cells in which it is contained (rods, cones or rhabdoms) and studying its properties *in vitro.* The second is the microscopic examination of the dissected retina. The third uses the principle of the ophthalmoscope. In the last method a beam of light is directed through

79

the pupil of the intact eye. After being reflected by the choroid and crossing the retina twice it emerges from the eye and can be studied (*reflectometry from the fundus of the eye*). In theory this last technique is the most direct, but it is difficult, and the interpretation of results uncertain. Conclusions drawn from microscopic and reflectometric studies will be dealt with later, after first considering the usual common indirect method.

The animal is sacrificed and the eye is dissected out. The retina is detached with forceps and placed in an aqueous solution of *digitonin*, a glucoside extracted from foxglove seeds. In this solution the membranes enclosing the external segments of the rods rupture, and the liberated pigments combine with the digitonin (whose molecular weight is 1,290) to give *micellae* of high molecular weight (about 270,000). Several pigment molecules become attached to one molecule of digitonin. This is not a true solution of pigment in water, but a colloidal suspension of micellae, which is contaminated with blood and retinal pigments (for example, xanthophyll from the macula). Contamination can be partially avoided if the retina is washed several times in a weak acid solution before extraction with digitonin. All these procedures must be carried out either in the dark, or with a dim red light ($\lambda > 700$ nm) in order to avoid decomposition of retinal pigments by light.

In another method known as the *flotation* method, the retinae are vigorously shaken in an aqueous sugar solution of the same density as the outer segments of the rods. The mixture is centrifuged to separate heavy debris and the resulting solution diluted and recentrifuged. The outer segments sediment to the bottom of the tube, and after washing can either be extracted with digitonin, or resuspended in a sugar solution of the same density. Theoretically, it would be more desirable to study the segments themselves rather than in solution with digitonin, but unfortunately suspensions of segments produce light scatter which hinders spectrophotometric measurement. The principle of difference spectra should always be used in the spectrophoto-metry of visual pigments, for two reasons. First, the impurities are eliminated, providing that they are stable to temperature variations and light. Second, the visual pigments are decomposed by light, and the absorption spectrum is changed. The difference spectrum between two solutions, the one intact and the other after exposure to light, allows the change to be identified. In fact the difference measured is that between the absorption of the pigment itself and that of the photoproducts which are formed as a result of photochemical action. The objection can be made that, as soon as spectrophotometric measurements are begun, light from the spectrophotometer begins bleaching the pigment, but in modern types of apparatus this effect is negligible. This can be verified by reading the spectrum a second time in the opposite direction, and confirming that for the same wavelength the results of the two measurements are practically the same.

If we consider as an example two extracts of human retina (Fig. 27), the one contaminated with blood (solid line curve, on which can be seen the α, β and λ oxyhaemoglobin absorption bands), the other having undergone further purification (dotted line curve). After bleaching by light, absorption by these extracts is shown by the discontinuous lines. The difference spectra, calculated and corrected in both cases to a maximum of 100, are both represented, within the degree of accuracy of the method by the dot-and dash line in the lower figure. It can be seen that the bleaching decreases all densities for $\lambda > 425$ nm with a maximum effect at about 500 nm, this decrease showing that the photoproduct absorbs less in this region of the spectrum than the original pigment. For $\lambda < 425$ nm the reverse is true and the photoproduct absorbs more than the original pigment. The radiation for which absorption is the same for both (in this case 425 nm), is known as the *isobestic point* and at this point bleaching has no effect.

When the extract being studied contains only one photosen-sitive pigment, the difference spectrum is the same, whatever

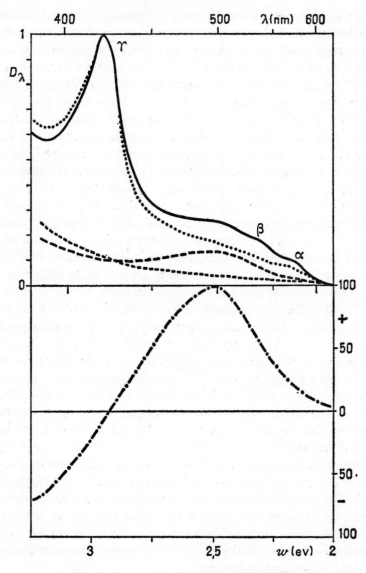

Fig. 27 Human retinal pigment, after Dartnall

light is used to bleach the pigment, and the result is related to the action spectrum of the pigment alone. For example in the case shown in Figure 27 rhodopsin is present practically alone in the extract, and the wavelength maximum, λ_m, of the action

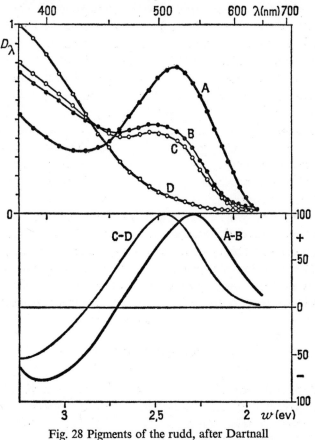

Fig. 28 Pigments of the rudd, after Dartnall

spectrum maximum, or photosensitivity, is probably about 500 nm. Frequently, however, the extract contains several photosensitive pigments, and in these cases Dartnall's partial bleaching method, devised in 1952, should be used. An extract of the rudd retina (Fig. 28) has in darkness the absorption curve A. After two hours' exposure to red light (660 nm) of a given intensity, a marked change is observed (B).

Irradiation for a further two hours with the same light source produces on the other hand only a slight change (C), demonstrating that practically all the red-sensitive pigment is bleached by the first exposure. If the extract is now exposed to white light, there is a marked change, indicating the presence of a second pigment upon which red light exerts no effect, although it is bleached by the short wavelength components of white light. The difference curves in the lower figure therefore distinguish the action spectra of two pigments whose λ_m are at about 545 and 510 nm respectively. In fact such a conclusion is only valid after a more complex procedure in which the two constituent pigments (one red-sensitive and the other not) are submitted separately to a homogeneity test based on the bleaching effects of various wavelengths.

This method works well where concentrations of pigments present in the extract are of the same order, or where the action spectra are sufficiently separated for a given radiation to act on virtually only one pigment. In each case, however, the extract should at least be suspected of containing more than one constituent in order to avoid jumping to hasty conclusions.

5.2. Rhodopsin

Rhodopsin has been studied more than any other visual pigment (especially in the frog and in cattle). The present state of knowledge on this subject is summarized below. It remains incomplete despite the large amount of work done in many laboratories, chiefly in England and the United States.

Since the beginning of last century, the absorption spectrum of rhodopsin has been measured, and shown to present an obvious similarity to the *scotopic* sensitivity of the human eye, determined in weak light after dark adaptation for twenty minutes. The rods constitute the organ of night vision, and the cones provide diurnal or *photopic* vision. There are many facts which provide evidence of this dual retinal function, amongst which may be mentioned the observation that the most sensitive region of the human retina to weak light levels lies between $10°$ and $15°$ from the centre of the fovea, and it is precisely there that anatomically the greatest rod density is found. The fovea contains only cones and is insensitive to very weak light. The astronomer Arago used to say that to see a weak star you must not look at it.

The dark-adaptation of rods permits vision in progressively weaker light as adaptation progresses and this at least partially is explained through the accumulation of rhodopsin in the outer segments of the rods. During the First World War it was observed that the night vision of soldiers who could not be relieved, and who ate only preserved food deficient in vitamins, deteriorated. In fact, the absence of vitamin A causes a partial or total loss of night vision. To try to combat this tablets of vitaminized chocolate were distributed to French soldiers at the beginning of the Second World War.

Vitamin A is an alcohol derived from the carotene $C_{40}H_{56}$, a hydrocarbon found in carrots. The retina contains various carotenoids, particularly carotene glycol ($C_{40}H_{56}O_2$), or xanthophyll, which constitutes the yellow macular pigment already mentioned. The carotenoids are unsaturated pigments, with alternating single and double bonds between the carbon atoms. Vitamin A is abundant in the retinal epithelial pigment, doubtless as a reserve for rhodopsin synthesis, but rhodopsin does not contain it in a free state.

When a retina is thoroughly dark-adapted, it has the pink colour which characterizes rhodopsin. Light causes the disappearance of this colour, which changes to yellow. Retinal photographs, or optograms, can be made by fixing the image with a solution of alum, which prevents further decomposition of the rhodopsin. When the retina has become completely yellow, a pigment known as *retinene* can be extracted, which has an absorption maximum of 385 nm when dissolved in chloroform. Retinene reacts with antimony trichloride to give a blue colour characteristic of carotenoids (absorption at at 664 nm). Prolonged exposure to light causes the yellow colour to disappear and vitamin A can then be extracted from the retina. Its maximum absorption in chloroform is at 328 nm and in antimony trichloride at 620 nm. Retinene can be chemically identified as vitamin A aldehyde ($C_{19}H_{27}CHO$) and therefore is also called *retinal*.

Besides retinene, one of the constituents of rhodopsin is definitely a protein. Its destruction by heat is similar to the coagulation of albumin, as noted by Ewald (1877). It also presents an ultraviolet absorption band at 278 nm, which is characteristic of proteins. The simplest approach is to consider rhodopsin as one of the *chromoproteins* formed by the combination of a protein, or *opsin*, and a *prosthetic group*. Opsin is the base and the prosthetic group the active element. Some chromoproteins contain a metal, iron in haemoglobin, the respiratory pigment in mammals and copper in the haemocyanin of fish. The visual pigments contain no metal; the prosthetic group of rhodopsin is retinene.

The absorption spectrum of cattle rhodopsin, for instance, (Fig. 29), shows three bands. One in the visible range ($\lambda_m = 498$ nm) serves to identify the pigment; a second in the near ultraviolet (340 nm) is less well defined, its maximum is only 20 per cent to 30 per cent of the height of the first. Finally, farther in the ultraviolet is a band characteristic of the aromatic amino acids of opsin. This band does not interfere with the photosensitivity, which can be verified by direct determination of the action spectrum (small circles in Fig. 29). The satisfactory agreement between the absorption and action spectra in the first two bands suggests a quantum efficiency Q independent of λ in this region.

After rhodopsin has been bleached by light, its spectrum is

modified, and the first two bands reappear as a single one (maximum 370 nm in this case) which characterizes the yellow photoproduct (dotted curve, Fig. 29). The position of the maximum depends on the pH of the solution.

A spectral curve similar to that of rhodopsin is found in all known visual pigments, the only difference being a translation

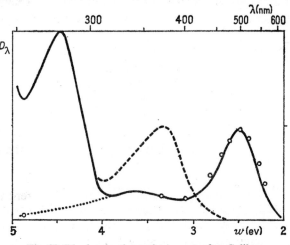

Fig. 29 Rhodopsin absorption curve, after Collins

parallel to the abscissa axis, provided that this is graduated in frequency or in fundamental photon energy (as in this book) and not in wavelength, which is another reason why this convention should be adopted. In particular the separation between the maximum λ_m of the absorption spectrum in the visible range and that in the near ultraviolet is always 1·2 eV, which shows that it is the same molecular structure, the prosthetic group, which produces this absorption.

How should the combination of opsin and retinene in rhodopsin be represented, and how does light act on its structure? To elucidate this problem, the important problem of *stereoisomerism* in retinene must be considered. This field has been investigated by the Harvard school (Wald and his collaborators) since 1951. Figure 30 represents the extended formula of retinene. It is shown with all the atoms of the conjugated carbon chain (with alternating single and double bonds) aligned in the same direction, i.e. the *trans* form. It is known that carbon atoms linked by a single bond can be rotated with respect to one another about this bond and that this is impossible with a double bond. Consequently rotation about the single bonds of the even-numbered carbon atoms (according to the numbering

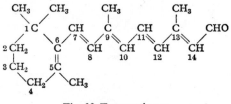

Fig. 30 Trans retinene

in Fig. 30) 6, 8, etc., give only one compound, although two compounds resulting from a 180° rotation about the double bonds of the uneven-numbered carbon atoms 7, 9, etc. One is the *trans* compound and the other the *cis* form, denoted by the number of the carbon in question. Since four carbons can be involved there are theoretically 2^4, that is, 16 different isomers (including the trans form) but their stabilities vary. The 8 forms, where carbon atom 7 is involved, do not occur because of *steric hindrance*, which results from the nearness of the methyl groups in the carbon chain to those in the ring. Among the 8 other isomers, those which involve a rotation about C_{11} also show steric hindrance, though to a lesser degree, due to the nearness of a methyl group and a hydrogen atom (Fig. 31). Two of these isomers are known, however (11-cis and 11-13-di-cis). Isomers 9 and 13 present no problem; these are the most stable, and are all known to exist since their synthesis has been carried out. The 11-cis isomer is present in natural rhodopsin. Artificial rhodopsins can be synthesized using other isomers, for instance the 'iso-pigment' formed from the 9-cis isomer, but at the present time have never been found in retinae.

When rhodopsin is bleached by light, the yellow photo-product obtained does not have the same absorption spectrum as 11-cis retinal which presents a λ_m at 376·5 nm as against 370

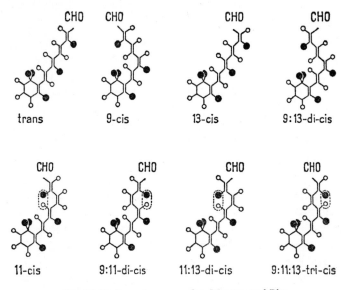

Fig. 31 Retinene isomers, after Morton and Pitt

for the photoproduct (at pH 9·2). The Liverpool school (Ball, Collins, Morton and Pitt), has shown that the photoproduct is a complex molecule formed from opsin and retinene, after elimination of H_2O through the combination of oxygen from an aldehyde group of retinene and hydrogen from an amine group in the protein. Thus in this case bonding is formed by the nitrogen atom. Is it the same for rhodopsin before irradiation?

This problem has not been solved. Besides this retinene-N-opsin bond, the possibility has been suggested of a retinene-S-opsin bond where sulphydryl groups would play a role as well as purely electronic supplementary bonds. The only certain fact seems to be that the prosthetic group bound to opsin consists of a single retinene molecule.

The chain of reasoning which leads to this conclusion is as follows. Direct measurement of the action spectrum gives the product $\varepsilon_w Q$, where ε_w denotes the photosensitivity, that is, the probability of absorption of a photon of fundamental energy w by the chromophore, and Q the quantum efficiency, i.e. the number of chromophore molecules transformed by the absorption of a photon. If the absorption spectrum of rhodopsin is determined under conditions such that the proportion of photons absorbed by the chromophore itself is measured (and absorption due to the protein and to photoproducts and impurities is eliminated) there is equality between the extinction coefficient α_λ and the photosensitivity ε_w, provided that in equation (16), page 30, the concentrations C are defined in molecules per unit length of the container. From measurement of the optical density D_λ, C can be deduced if the mass of rhodopsin present in the cell is known, and if the value of Q and the number P of chromophores in the rhodopsin molecule are known. These measurements are difficult, since digitonin and impurities must be eliminated in order to estimate the amount of rhodopsin alone in the solution. The best measurements made by this technique give the value $22,500\,pQ$ as the molecular mass of rhodopsin.

On the other hand, this molecular mass can be estimated by a quite different method, using the sedimentation constant in the ultracentrifuge. The micellae of rhodopsin and digitonin correspond to a molecular weight of from 260,000 to 290,000. The proportion of opsin in the micella can be calculated, by measuring the nitrogen, about 14 per cent, which leaves 36,000 to 41,000 for the molecular mass of rhodopsin. The product pQ is therefore of the order of 0·58.

The number p is a whole number and its minimum value (unity) corresponds to a quantum efficiency which would be of the order of 0·6 under experimental conditions. Thus, under these conditions, is should be possible to bleach a maximum of half of the molecules of visual purple. Whence the conclusion $Q = \sim 0.5, p = 1$.

At the present time it is believed that the first effect of the absorption of a photon by rhodopsin is the isomerization of the 11-cis to the trans form. This would be the only photochemical reaction, with a quantum efficiency of unity. The later reduction of the quantum efficiency to a value Q of the order of 0·5 is due to secondary reactions which cause the reaction to reverse by about half by a kind of immediate regeneration. (In this context one notes how little significance the concept of quantum efficiency has and how it should be handled cautiously. One is inclined to think at the present time that in the overall process of rhodopsin bleaching, Q may in fact have any value between unity, for very weak light, and zero, the limiting value for very intense light). Since the trans form is more stable than the cis form, photo-isomerization releases energy. Photon energy is the most important factor in photosynthesis in green plants. The sun's radiation is captured by chlorophyll and complex molecules so built up from simpler ones. Vision, however, proceeds in the opposite direction and the function of light is only to trigger off a process which is theoretically possible without a supply of external energy. It is even probable that the choice of the relatively unstable 11-cis isomer facilitates the catalytic effect of the photon. Photo-isomerization is a process which can be studied in the laboratory on much less complicated structures than rhodopsin; for example, the reversible transformation of fumaric acid to malic acid, or of nitrous acid at low temperature (20° K), under the influence of infrared photons of only 0·4 eV, with a quantum efficiency of unity.

The primary isomerization reaction is followed by complex secondary reactions which the Harvard school has tried to follow by the absorption spectra after slowing down the reactions by lowering the temperature, and so obtaining a whole series of products which have been called prelumirhodopsin (primary trans phase), lumirhodopsin, meta-rhodopsins I and II, leading finally to the complex of trans-retinene and opsin. With intense irradiation a colourless mixture of opsin and trans vitamin A is obtained. In vertebrates all the intermediate products have very short lives at normal eye temperature and their exact structure, if not their existence, remains conjectural. Experiments using a very brief flash of high intensity suggest that other isomers can be produced as intermediary products (particularly the isomers in the top line in figure 31, whose stability is intermediate between the 11-cis and the trans forms). It is possible that these are the effects of multiple absorption by the same rhodopsin molecule, such effects being of negligible probability under the conditions of normal vision.

Between the photochemical reactions in the outer segment of the rod, and the nervous impulse, which will be described in the next chapter, there is an amplification mechanism, the nature of which remains unknown. Borting and Bangham (1966) believe that there is movement of ions across a hole in the visual cell membrane, this hole being a result of the photochemical reaction.

The process by which rhodopsin is regenerated in the dark has been thought for a long time to have two components. There is rapid regeneration in the rod itself, the yellow photo-product being transformed into rhodopsin, which assumes that an inverse trans-cis isomerization takes place with the required energy being produced by an enzyme secreted by the inner segment. Then there is a slow regeneration, from vitamin A stored in the epithelium, which of course also requires an enzyme. Although in theory these processes are reversible and therefore should need no supply of new products, a fraction of the substances must be lost at each regeneration, so that the reserves would become exhausted, and haemeralopia would occur if fresh vitamin A from food were not supplied through the circulation.

The enzymic mechanisms of regeneration are almost unknown.

It is supposed that the reduction of vitamin A in retinene is due to coenzyme DPN (diphosphopyridine nucleotide). Adenosine triphosphatase (ATPase) appears also to exist in the membranes of the outer segments of the rods, in larger amounts after illumination of the retina than when it is dark adapted. The mitochondria of the inner segment also need to be investigated. Although rhodopsin has been known for nearly a century its actions have not fully been elucidated. Rhodopsin can be studied not only by digitonin extraction, but also in the outer segments of the rods themselves suspended in sugar solution. The outer segments are easily detached from the inner segments because of the fragile constriction which separates them (see Fig. 26, p. 77). The considerable extent to which light is scattered by these suspensions has to be taken into account, and light scattered in a forward direction, that is, in the same direction as the regularly transmitted light, is collected and passed on to the photoelectric receptor. The retroscattered fraction cannot be recovered, and it must be accounted for by calculation, which is rather unreliable.

In spite of these difficulties, there is surprising agreement between the absorption spectra of frog rhodopsin extracted but not irradiated and rod suspensions. This shows that the combination of rhodopsin and digitonin to form micellae does not alter the absorption spectrum. On the other hand, the difference spectra (between irradiated and non-irradiated rhodopsin) show a small deviation (of from 3 to 5 nm) in the λ_m of the difference spectrum, which is slightly displaced towards the longer wavelengths for the suspension compared to the extract. This can be explained by the presence of a supplementary photoproduct in the first case, whose λ_m is between 460 and 470 nm.

There is also an absorption difference due to the way that the rhodopsin molecules are orientated within the rods. Their long axes, or at least those of their prosthetic groups, are parallel, and this plane is perpendicular to the axis of the inner segment, judging from the fluorescence observed in polarized light. The floating segments are, on the other hand, randomly orientated, which re-establishes isotropy. It is possible that in the micella there is also some alignment of the rhodopsin molecules. It is difficult to describe the geometric structures more fully since these are uncertain. In spite of this many attempts have been made to draw up schemes, some very detailed, which describe the alignment *in situ* of retinene and opsin in the rods. Some are reminiscent of comic strip cartoons, where cis-retinene is represented as being delicately enveloped in an alveolus of opsin, to awaken at the shock of the photon, stretch from cis to trans, and, thus compelled to leave its bed with regret, and so initiate vision.

5.3. Visual pigments in animals

The first comparative research on visual pigments in animals was attempted by Köttgen and Abelsdorff (1896) on four mammalian, one bird, three amphibian and eight fish species. There appeared to be two distinct types of pigment, one, of the rhodopsin type, with a λ_m in the blue-green at about 500 nm and the other of the porphyropsin type with a λ_m in the yellow-green, at about 540 nm. The first type was found in all species except fish, while the latter was only found in fish.

Working on perch retina, Wald (1937) showed that a retinene which differed from that of the frog could be extracted with chloroform. Its absorption maximum was 405 nm (as against 385) and on reaction with $SbCl_3$ it gave a λ_m of 703 nm. In addition the completely bleached retina yielded a different vitamin A, whose λ_m were 355 and 696 nm with the same reagents. These substances are known as retinene$_2$ and vitamin A$_2$, to distinguish them from retinene$_1$ and vitamin A$_1$ which occur in rhodopsin. It has finally been established that their only difference is an additional double bond in the ring, between carbon atoms 3 and 4 (see Fig. 30, p. 88) with the loss of 2 hydrogen atoms. This supplementary bond displaces all the spectra by about 20 nm towards the red compared with those of retinene$_1$.

Recent studies have demonstrated that, at least in vertebrates, all visual pigments have a similar structure to that of rhodopsin, and are formed by combination of retinene$_1$ or retinene$_2$ with an opsin characteristic of the species under consideration. Opsin variation accounts for modifications of the absorption spectrum in different animals. This remarkable unity within the animal kingdom is exemplified by the following experimental observation; if the absorption spectra of pigments are drawn with frequency (or fundamental photon energy w) as linear scale on the abscissa, all the curves can be superimposed if their maximum λ_m are displaced to the same point. This is shown in the positive part of figure 32, for 4 pigments whose λ_m range from 467 to 562 nm and whose difference spectra (as percentage maximum) are represented as a function of the variation Δ_w of the fundamental photon energy corresponding to λw. Consequently only λ_m need be known in order to characterize a pigment. A nomogram, once established, can be used to re-establish the entire absorption curve for any wavelength and therefore the action spectrum if the quantum efficiency remains constant (Dartnall). For greater accuracy it is better to construct two nomograms, one for pigments with a retinene$_1$ base, the other for those with a retinene$_2$ base (in anticipation of the possible discovery of a retinene$_3$?). The question has been raised of what happens if an animal whose pigment is based on retinene$_1$ is depleted of vitamin A_1 and given vitamin A_2. For albino rats, it seems that conversion of vitamin A_2 to retinene$_1$, occurs in the retina, leading to the formation of normal rhodopsin, but a small amount of an unusual 517_2 pigment is also produced. We use here the simplest method by giving the value of λ_m in nanometers, indicating whether the pigment has a retinene$_1$ or a retinene$_2$ base by a subscript. To be sure of the presence of a pigment in a retina it is not enough to make a retinal digitonin extract and simply read off a result on a spectrophotometer. A homogeneity test must be applied, to check whether or not there is a mixture

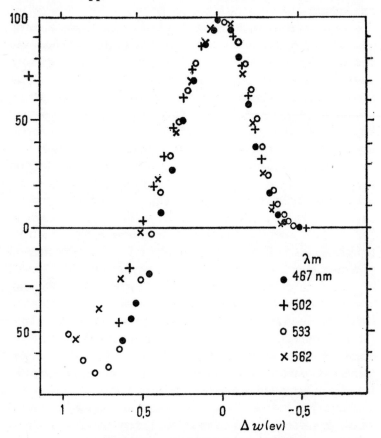

Fig. 32 Various pigments, after Dartnall

of visual pigments in the extract, and to correct for absorption due to impurities. This can be done using the difference spectra. The pigment can be bleached in the presence of hydroxylamine which reacts with the photoproduct to form an oxime of retinene whose absorption is in the UV and which therefore interferes much less with absorption by the pigment than does the photoproduct itself.

Prior indication as to whether the pigment has a $retinene_1$ or a $retinene_2$ base is indicated by the λ_m of the photoproduct (retinene + protein) which is situated at about 380 or 400 nm respectively. The results are more certain if the bleaching is taken as far as the vitamin A phase. The chloroform extract is then added to antimony trichloride, and the blue product obtained has a λ_m at about 620 nm for vitamin A_1 and at about 693 nm for vitamin A_2.

The variation of λ_m with the opsin, from one animal species to another, is more marked than the variation seen when $retinene_2$ is substituted for $retinene_1$ in a pigment with the same opsin. The rule that $retinene_2$ pigments show a shift of absorption into the long wavelengths by comparison with $retinene_1$ pigments is not an absolute one as exemplified by pigment 523_2 in the carp, and pigment 528_1 in the gecko.

The main results on rod pigments in the descending evolutionary scale (if one can so describe it!) will now be reviewed.

Mammalian pigments all have a $retinene_1$ base. In man there is some uncertainty between Crescitelli and Dartnall's value of 497 and Wald and Brown's of 493. This will be discussed later. The chimpanzee with 491 is furthest into the short wavelengths, and the rabbit with 502 into the long. The grouping is thus very compact. Birds also have $retinene_1$-based pigments, falling within an even narrower range, between 500 and 503 nm.

Among the reptiles, the rattlesnake and the alligator fall at 500; geckos show considerable variety between 516 and 528 nm; these are all $retinene_1$-based pigments. It appears that some species may have a second pigment with a λ_m of short wavelength; for example *Oedura monilis* shows a mixture of 85 per cent of 518_1, and 15 per cent of 457_1. The biological significance of such results is doubtful. According to Walls (1942), the cone is the primitive type of visual cell, rods being developed later in the course of evolution, in relation to the transition from

97

diurnal to nocturnal life. Nocturnal geckos with rods would retain a souvenir of their diurnal ancestors, which would explain the relatively long wavelengths of their λ_m and the trace appearance of pigments with a maximum in the blue during evolution would indicate a better nocturnal adaptation.

In adult amphibians, most species have retinene$_1$-based pigments with a λ_m between 500 and 505 nm, but in some aquatic species retinene$_2$ pigments appear, for example in the South African toad *Xenopus laevis*, which has a mixture of 8 per cent of 502 with 92 per cent of 523. Mixtures of the same kind are confirmed in many larvae with the retinene$_2$-based pigment predominating. At metamorphosis this disappears, leaving only a retinene$_1$ pigment. It is possible that this is related to the transition from aquatic to terrestrial life.

Fish show a remarkable variety of pigments, which are often multiple in any one species. It was for a long time assumed that sea-water fish possessed retinene$_1$ pigments, fresh-water fish retinene$_2$ pigments, and euryhaline species a mixture. These simplified views have been abandoned, but it is true that there is a remarkable correlation in fish between their pigments and the luminosity of their environment.

At great ocean depths, daylight, which reaches 500 m, is concentrated in a narrow blue band of about 480 nm, the region of greatest water transparency. Deep-sea fish have retinene$_1$ pigments whose λ_m varies between 478 and 490 nm. They are therefore specially suited for picking up what little of the sun's radiation filters down to great depths. Besides this, luminescent marine animals often have a maximum emission in this same spectral region, and possibly the vision and the light emitted by a species are mutually adapted (see Chapter IX). As their habitat nears the surface, the λ_m of marine fish approaches 500 nm, which indicates a compromise between the transparency of the water, which plays a less selective part in shallow water, and the relative number of photons in sunlight, which is maximum in the infrared (inasmuch as this idea has any meaning, since it depends, as has been stated already, on the convention chosen for the spectral scale.) Some *Labridae* species have retinene$_2$, based pigments, although they are entirely marine.

In fresh-water fish, although retinene$_2$ predominates, pigment mixtures containing retinene$_1$ are found, for example in the ablet ($510_1 - 553_2 - 550_2$), and the bullhead ($510_1 - 543_2$). The transparency of fresh water is shifted towards the long wavelengths compared with sea water.

Some of the fish which can accommodate to great variations in salinity, e.g. *Gillicthys mirabilis*, have a single pigment 512_1, midway between values for sea and fresh water. Others have retinene$_2$-based pigments. The most interesting example is that of migratory fish which spend their early life in rivers, and later in the sea, or *vice versa*. Some *Salmonidae* show a mixture in practically equal proportions of 507_1 and 533_2. An interesting case is that of two similar species of *Salmonidae* of the genus *Salvelinus* of which one has pigments 503_1 and 527_2, the other 512_1 and 545_2. The hybrids obtained by crossing have the four pigments in their retinae. The Lake Michigan lamprey has only one pigment (497_1), like that caught in the sea (518_2). This seems an exception to the rule, but perhaps may be explained by the fact that the first of these fish migrates to the sea where it spends its adult life, and the second returns to fresh water to spawn. In this case it is not the environment which is concerned but the migratory hormone, which modifies the internal *milieu* in advance. It must be the same for the adult eel which, while it lives in river water, has the mixture $502_1 + 523_2$, and when it goes to spawn in the Sargasso sea, and turns silver, its retina contains only 487_1.

Modifications with environment have been studied in detail on the fresh-water fish *Scardinius erythrophthalamus* which has a mixture of 510_1 and 543_2. Fish caught in the winter have often only 15 per cent of pigment 510_1 while in the summer this proportion may reach 85 per cent (Fig. 33). But this is not a natural annual rhythm, because if fish are caught in winter from a well-lit aquarium, the amount of 510_1 is considerably increased (dotted curves, white circles) although it can be lowered in summer by keeping the fish in the dark (black circles). The reason for these natural changes, which are synchronized with day length, is not clear. Dartnall thinks that these modifications may be linked to recognition of the seasonal food supply and a mate.

In conclusion, one can say in a general way that the marked predominance of retinene$_1$ in the animal kingdom is due to

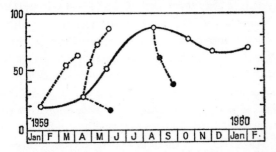

Fig. 33 Pigment variation in *Scardinius erythrophthalamus*, after Dartnall

vitamin A$_1$ being more widespread, and probably simpler to synthesize. The change of opsin is sufficient to put the λ_m in better agreement with the animal's luminous environment. It is only in the case where a λ_m of long wavelength is necessary (as in some fresh waters) that the animal has retinene$_2$ since it would be difficult to find an opsin which would give the same result with retinene$_1$. Another case is that of animals in which a change in λ_m is useful during the life of the individual animal. It is then easier to add a double bond to retinene$_1$ than to change the opsin. To put an end to these teleological considerations, it is expedient to put the reader on guard against specious reasoning which springs from the approximate coincidence between a λ_m in the absorption spectrum (which really exists) and the maximum in the sun's spectrum which is purely a convention, and in any case would be in the infrared if the number of photons, the only quantity of interest to photobiology, is taken into account.

All the pigments which have been mentioned here relate (it is thought) to the rods of vertebrates. Our knowledge about the cones is practically nil, since up to now no method has been found of extracting any pigments they may contain. In the grey squirrel, which has only cones in its retina, Dartnall has shown a pigment 502_1. In the chicken, which has eight times as many cones as rods, Wald has described a pigment 562_1 which he called *iodopsin*, and in the pigeon Bridges has isolated a pigment 544_1. In fact there is no proof that these are really cone pigments; this question will be discussed in Chapter VII.

Finally, to close this review of visual pigments in animals, mention must be made of the invertebrates. Our knowledge is still rudimentary. Visual pigments have been extracted from cephalopods, and their constitution seems to be analogous to that of rhodopsin, with λ_m at 475_1 in the octopus, 492_1 in the cuttlefish and 493_1 in the squid. However, the method of decomposition of light differs from that in vertebrates; for example the metarhodopsin of the octopus is stable and in equilibrium with rhodopsin under the action of light which modifies the 11-cis form of retinene$_1$ to the trans form. Because of this fact, it is not strictly correct to speak of the bleaching of rhodopsin by the absorption of photons, but only of stereo-isomerization.

In crustaceans, the lobster has a pigment 515_1 which gives on irradiation a stable metarhodopsin ($\lambda_m = 490$ nm) as in the octopus. In the crab *Limulus*, on the other hand, the pigment 520_1 is bleached like the rhodopsin of vertebrates. A fresh-water crustacean, the crayfish *Orconectes virilis*, has two pigments. One (508_1), is rapidly bleached by light to retinene and opsin, although the other (562_1), ends as a relatively stable meta form. It is interesting to note that this pigment, displaced towards the long wavelengths, behaves a little like known pigments with a retinene$_2$ base found in fresh-water fish.

101

In insects, the only pigment which has been isolated to date is that of the bee (440_1) which shows two pecularities. The λ_m is of shorter wavelength than any other pigment and it is soluble in water and therefore does not need to be extracted with digitonin to be studied. These results are, however, extremely sparse when it is considered that the species of insects are more numerous than all other living species, both animal and plant.

REFERENCES

BORTING, A., and BANGHAM, B., Exper. Eye Res. **6**, 400 (1967)

BRIDGES, C. D. B., Vision Res., **2**, 125 (1962)

COLLINS, F. D., LOVE, R. M., and MORTON, R. A., Biochem. J. **51**, 292 (1952)

CRESCITELLI, F. and DARTNALL, H. J. A., Nature, **172**, 195 (1953)

DARTNALL, H. J. A., *The Visual Pigments*, London, Methuen (1957)

DARTNALL, H. J. A., in DAVSON, ed., *The Eye*, **2**, 321, N.Y., and London Academic Press (1962)

MORTON, R. A. and PITT, G. A. J., Fortschr. Chem. org. Naturst. **14**, 244 (1957)

WALD, G., Nature, **139**, 1017 (1937)

WALD, G., Science, **162**, 230 (1968)

WALD, G. and BROWN, P. K., Science, **127**, 222 (1968)

WALLS, G. L., *The Vertebrate Eye and its Adaptive Radiation*, N.Y. and London, Hafner, 1963

CHAPTER VI

Visual Responses

6.1. Electrophysiology

Light absorption by the pigment of a rhabdom, rod or cone constitutes the initial phenomenon of vision, and determines its photosensitivity, simply by the probability of absorption of a photon in the chromophore of a pigment. From this point of view all photons are equivalent and if, for example, the light sensitivity of an animal with rhodopsin is maximal in the blue at about 500 nm, this is simply because the photons whose elementary energy w has the value of 2·48 eV (see equation (3), page 4) have most chance of being captured by the rhodopsin in the rods. But from the moment a photon is absorbed, its action is the same, the 'colour' of the photon is not 'seen'. Obviously, for this to be so, absorption by the pigment must be due to the chromophore, and not to the protein to which it is linked. It is also necessary that the quantum efficiency should not fall to zero. It appears to remain constant in the visible range, but obviously for photons with weak w which are unable even to cause the isomerization of retinene, there will be absorption by the chromophore without photochemical result, producing only a small rise in temperature. In such cases Q is zero.

Following the complicated and little understood secondary reactions set off by the absorption of a photon in the visual cell, a visual impulse is initiated and propagated to the brain. It is then transmitted as a motor impulse governing the appropriate reactions. The animal gets in this way most of the information supplied to it about the objects which surround it. The essential features of the visual impulse, which have been derived largely by the use of electrophysiology, will now be described.

It is convenient to start at the lower end of the animal scale, where the situation is simpler than in the extremely complex retina of vertebrates. Bivalve molluscs, for example, react to light and shade, and it has been shown that the photosensitive receptors are near the siphon. The potential differences which arise in these receptors following light stimulus can be recorded with fine electrodes and amplified electronically. Three different types of response can be obtained, depending on the cell from which the potential difference is picked up. First, the *on* type, where the response consists of repetitive discharges which continue during light stimulation and cease when the light is

Fig. 34 Response of the mollusc *Spisula* to 600 nm, after Kennedy

switched off. Second, the *off* type, where nothing is recorded during illumination, and switching off the light causes a volley of discharges which gradually slows down and stops. Finally the *on-off* type, where there is a response at the beginning and end of illumination. In some cases there is a spontaneous discharge in the dark, which is inhibited by light, and whose frequency is increased by a marked off response when the light is extinguished (Fig. 34). It will be noted that the electrical response consists of very brief potential waves of virtually the same height (several millivolts are recorded between the searching electrode and a neutral electrode in the mass of the organism), the only variable parameter being the frequency. The nervous impulse may be compared to a sequence of 'dots' in morse, the higher or lower frequency of these *spikes* being the only language in which the message can be coded. The inhibition of the response by light and its augmentation by darkness are obviously especially useful in marine molluscs, which live in a low-level constant luminosity, where a shadow is a sign of approaching danger, and provokes a defence reaction. In the sea urchin *Diadema*, it has been confirmed that movements of the spines, caused by a shadow falling on some part of the body, are inhibited by light on another region, which indicates nervous interaction between receptors.

In cephalopods, it has recently become possible to interpret the nervous mechanisms involved after the absorption of photons by rhodopsin. The eyes of these animals favour such an analysis, since they are composed of a group of identical rhabdoms, whose visual cells end in *axons*, or nerve fibres, which relay directly to the optic lobes in the brain through perforations at the back of the eye. If two micro-electrodes are inserted into adjacent visual cells it is found that the region illuminated by a very localized light point (40 μ) becomes negative. The potential is continuous and graduated, increasing with the light intensity. This is known as an *S potential* (slow). It appears to be due to a variation in the permeability of the membrane of the illuminated cell, resulting in depolarization (decrease in resting potential), due to ions of all types leaving the cell. This depolarization sets off quite a different phenomenon in the axon. which is related to its permeability to Na^+ ions only. This action potential is the *all-or-none* phenomenon previously described, that is a volley of spikes which is propagated along the length of the axon. These basic potentials have the same amplitude and variable frequency.

In crustaceans, the compound eye of the crab *Limulus* has been extensively studied by the electrophysiologist Hartline for almost 40 years. The ommatidium has a characteristic structure; it contains 10 to 20 photosensitive cells (*retinula cells*), in contact with an *excentric cell*, from which the message is transmitted by the axon. It is probable that the retinula cells are the source of an S potential whereas the excentric cell gives

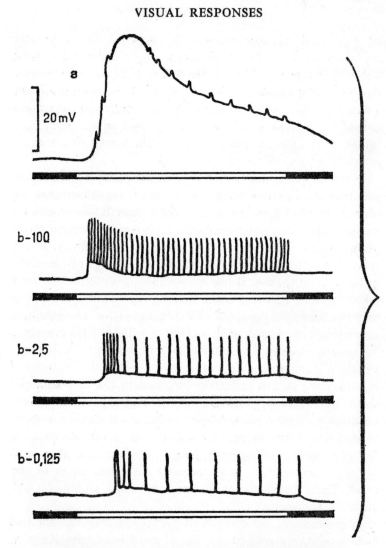

Fig. 35 Response of the ommatidium of the crab *Limulus*
(a) micro-electrode in the retinula, after Tomita: (b) micro-electrode in
the excentric cell at various light intensities, after Fuortes. Time
in abscissa, recorded potential in ordinate. The white segment
represents the duration of light excitation (here 1·7 seconds).

106

rise to an action potential. It is, however, difficult to isolate them, and, the responses in fact show a mixture of both (Fig. 35). The variation in spike frequency with the light intensity is shown and also the variation in *latency* (delay in the beginning of the response). It is not known whether transmission between the retinula cells and the excentric cell is chemical or electrical.

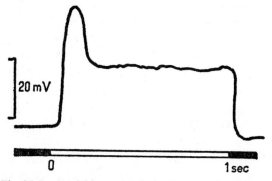

Fig. 36 S potential in worker bee, after Naka and Eguchi

The interactions between ommatidia are very important in *Limulus*, and will be discussed later.

In insects, there is no excentric cell, and the visual impulse is relayed through a synapse (nerve junction) with a ganglion cell, on leaving the ommatidium. The situation is similar to that seen in *Limulus*, but the S potential has a characteristic wave form. There is an initial marked maximum, and no spikes appear on the record (Fig. 36). When the responses of the visual cells of an insect are recorded with microelectrodes in an ommatidium (difficult but possible), some ommatidia are seen to be sensitive to the polarization of light and others not. This throws doubt on the theory based on the fine structure of the rhabdomeres, previously described. An additional difficulty is that in *Calliphora* the discrimination of polarization is better for short wavelengths than for long which is in favour of the light guide theory.

In vertebrates, visual electrophysiology is more than a century old. The *resting potential* of several millivolts between the cornea and the back of the eye was discovered in 1849 by Du Bois-Reymond, and the superimposed *action potential* due to light by Holmgren in 1865. For a long time the technique of recording *electroretinograms* (ERG) consisted in studying the potential difference between large electrodes placed one against the cornea, the other on the forehead. Many spurious effects are recorded by this method (movements of the eyes or eyelids, contraction of the iris) as well as changes resulting from illumination of the retina. These are extremely complex and light diffused outside the so-called retinal image by the heterogeneity of the lens, the vitreous body and the retina itself is largely responsible. This is proved by the fact that an ERG is recorded even when the image falls on the blind spot (where the optic nerve enters the eye) where the retina is not photosensitive. The ERG originates chiefly in the rods and the absorption curve of rhodopsin can be plotted exactly from the action spectrum obtained.

In spite of medical interest in the ERG, further consideration will be left aside and intraretinal recordings described. These can be more directly interpreted, but the technique is difficult. Microelectrodes can now be made with points less than $1\ \mu$ in diameter and these can be inserted into a single retinal cell to record its electrical response. One difficulty consists in knowing where it is inserted at any given moment. Fortunately a sudden variation is detected at the moment when the microelectrode crosses the membrane which separates the choroid from the epithelial pigment of the retina, which helps to localize the microelectrode.

Very varied electrical responses are recorded depending on the region where the microelectrode is inserted. In the visual cells themselves, there is a potential called the ERP (*early receptor potential*) with an extremely short latency, that is, there is almost no delay between the absorption of photons by the cone or rod and the appearance of the potential. The ERP is composite and seems to be the result of at least two phenomena, one of maximum amplitude at the junction of the inner and outer segments of the receptor, the other probably linked to the epithelial pigment. This latter effect seems to be without a visual role, although the former, whose amplitude is proportional to the amount of pigment bleached by the light flash, is apparently the true response of the outer part of the receptor, and gives rise to a potential in the inner part. This has been recently discovered by Tomita and will be described later.

In fish, Svaetichin discovered an S potential which arises at the level of the bipolar cells. This is a slow, stable, graduated response, which remains constant during a constant illumination, unlike most retinal potentials, which vary or at least change in polarity during constant illumination. The action spectrum of this S potential, the spikes recorded in the ganglion cells and neuronal responses throughout the nerve circuit carrying the impulse to the brain, will be dealt with in the next chapter in connection with colour vision. In fact, in vertebrates—and in man in particular—there are four synapses where the nerve impulse passes from one neurone to the next in the chain of visual information. The first is between the cone (or rod) and a bipolar cell. The second is between a bipolar cell and a ganglion cell. The third is between an optic nerve fibre (axon of a ganglion cell) and a neuron of the *lateral geniculate nucleus* (LGN) which is an important relay at the entry to the brain, and finally the fourth is between the fibres leaving the LGN and the cells of the striated zone of the cortex where the visual impulse finally arrives. The phenomena occurring at these various levels will be dealt with later.

6.2. Luminous thresholds

When a given surface S of an animal's retina receives a monochromatic energy W during a time t, either no impulse is transmitted along the optic nerve, and there is no reaction, or else information is supplied to the brain and the animal can react. The limiting case separating the two possibilities is known as the threshold, which is thus the weakest light which can elicit a response.

In adult man, the response is found by a simple subjective method, the subject says 'I see' or 'I do not see'. These alternatives can be differentiated by suitable statistical treatment (since there are cases where the subject is at fault, and claims to see something which does not exist) and the threshold defined by a 50 per cent probability of perception.

The laws of the threshold to light are complicated. If t is very small (less that 0·01 second for instance), W measures the threshold. On the other hand, for prolonged exposure (several seconds) it is measured in terms of the energy flux, i.e. the power W/t received by the surface S. In the first case there is complete temporal summation, and the law of reciprocity comes into play. Clearly there is a connecting region where W and t act independently.

In the same manner, when S is very small (apparent diameter less than an angle of one minute), there is total summation, which is in this case spatial and only W is concerned. For large surfaces, of several degrees of apparent diameter, it is effectively the quotient W/S which is the variable concerned; the flux per unit surface of the retina is the quantity which governs the threshold. Here again an intermediate region exists where summation is partial, W and S functioning independently.

The retina has to be considered as an extremely heterogeneous receptor surface and the position on the retina of the excited surface S must be specified whether for example it is in the fovea or at a given angular distance from it.

Finally, the prior condition of the visual system must be known. When walking from full sunlight into a dark room such as a cellar, at first nothing is seen, then, after some minutes the eye becomes accustomed, there is dark *adaptation*. After about 20 minutes, practically no further change in threshold takes place. Figure 37 represents typical adaptation curves as a function of time spent in the dark. $\log_{10} W$ is taken as ordinate (W being the energy which strikes the retina at threshold), so that the progress of adaptation can be followed for great variations of W. A break in the curve appears after several minutes: this is interpreted as the transfer from cone vision to

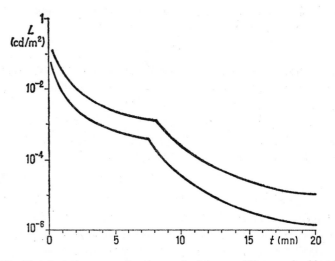

Fig. 37 Adaptation curves (maxima and minima on 100 normal subjects)

rod vision; rods adapt better and are therefore finally more sensitive. In the fovea, which contains only cones, only the first phase of adaptation is seen. The same phenomenon is observed if W is supplied by a red-light source which excites only the cones.

If the threshold is determined as a function of the wavelength λ of the monochromatic light used to excite the retina, it is obvious that W varies considerably. The energy required to reach the threshold is at a minimum in the centre of the visible spectrum and increases considerably towards the infrared or ultraviolet. For the rods, the *scotopic luminous efficiency* is defined by the statement:

$$V'_\lambda \Phi = \text{constant} \tag{20}$$

Φ denoting the energy flux which strikes the retina and corresponds to the threshold. Quite obviously V'_λ is only defined in relative value; $V'_\lambda = 1$ is taken for the wavelength corresponding to the minimum Φ.

It is clear that the energy flux Φ has no physiological value. On the contrary the quantity:

$$F' = K' V'_\lambda \Phi \tag{21}$$

has (K' denotes a constant), since at the threshold from equation (20), the *scotopic luminous flux* F' has a constant value at any wavelength. If several wavelengths are mixed, it is accepted (Abney's law) that the threshold is effectively reached when the sum F' of the fluxes reaches a constant value (20), whence the usual convention of defining the scotopic luminous flux:

$$F' = K' \int V'_\lambda \times d\Phi \tag{22}$$

$d\Phi$ representing the elementary energy flux carried by the monochromatic radiation between wavelengths λ and $\lambda + d\lambda$. In principle the integral has 0 and ∞ as limits but since V'_λ vanishes outside the visible spectrum, the integral can be limited to the visible spectrum.

It is fairly simple to determine the spectral sensitivity V'_λ of the rods near the threshold, since the cones are practically blind under these conditions, except at long wavelengths, when the retina is thoroughly dark adapted, and when an area of about $10°$ from the fovea where the rod density is at a maximum is used. It is much more difficult for the cones. The fovea has to be used, but the threshold is difficult to determine. Questions more technical than physiological will be left aside; the *photopic luminous flux* F can then be defined by the convention:

$$F = K \int V_\lambda \times d\Phi \tag{23}$$

V_λ being in this case the photopic luminous efficiency. The curves V_λ and V'_λ are shown in Figure 38. The maxima (for man) are respectively at about 555 and 507 nm, if the abscissa is graduated in λ or in w (not important since it is concerned with energy ratios, by definition). The displacement towards the

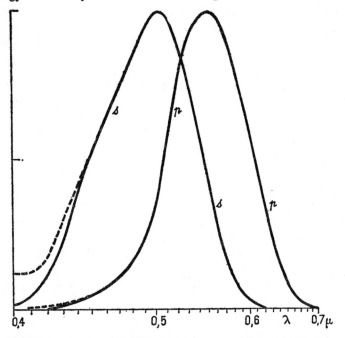

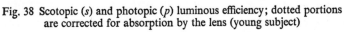

Fig. 38 Scotopic (*s*) and photopic (*p*) luminous efficiency; dotted portions are corrected for absorption by the lens (young subject)

blue when the eye is dark adapted past the stage of photopic vision (to the left of the angular point of the break, Fig. 37, see p. 111), to scotopic vision (to the right of this point), is known as the *Purkinje phenomenon* in honour of the Czech physiologist who described it nearly a century and a half ago.

Beginning with the luminous flux, photometric quantities can be deduced by simple geometric relations. The *luminous intensity* of a source of small dimensions in a given direction is obtained by the quotient of the flux F' emitted over a small solid angle

113

around this direction, by the value Ω of the angle. It is assumed, of course, that the light is propagated in a homogeneous and non-absorbent medium and that, due to this fact, the energy contained by the solid angle remains constant by virtue of the principle of rectilinear propagation. If the light source has an appreciable apparent dimension, it is no longer intensity but *luminance* which is used in the photometric definition, this is in the direction under consideration, the quotient of the intensity emitted by a small surface S of the source, divided by the value of S projected on to a plane perpendicular to the direction of observation. Finally, when a light flux F reaches a surface S, the quotient F/S is known as the *illumination* of the surface.

To define the photometric values unambiguously two functions V_λ and V'_λ are selected. Variations are observed from one subject to another, in part due to absorption by the lens and macula. The International Commission on Illumination (C.I.E.) has established universally accepted V_λ and V'_λ tables for a fictitious being called the reference observer (phototopic or scotopic). The observer represents the average subject of less than 30 years of age. It remains only to adopt values for K and K'. For this it has been agreed that the luminance of a black body at the temperature of solidification of platinum (2,042° K) has the value of 60 in both photopic and scotopic units. The choice of this number results from the fact that the luminous intensity defined in this way (*candela*) is close to the old empirical unit of candlepower. Other units are as follows: *lumen* for the flux (1 candela in one steradian), candela per square metre for the luminance and lumens per square metre (lux) for the illumination.

From these conventions the numerical values of K and K' are respectively 680 photopic lumens per watt, and 1,750 scotopic lumens per watt. A monochromatic source with $\lambda = 555$ nm therefore radiates 680 photopic lumens per watt, and any other source radiates less. Similarly a source with $\lambda = 507$ nm radiates 1,750 scotopic lumens per watt. To give an idea of the order of numerical values, the sun at midday in summer illuminates the earth with about 100,000 lux and a white matt horizontal surface has a luminosity of about 30,000 candelas/m^2. At these high levels photopic quantities are concerned, since the cones are the only receptors active in strong light. The full moon illuminates the earth with about 0·2 lux, and the luminance of a white matt horizontal surface is about 0·06 candela/m^2. The cones can still see (their threshold is of the order of 10^{-3} candela per square metre) but the rods are definitely also functioning, whence an ambiguity; is vision photopic or scotopic at this point? The question is avoided by postulating a *mesopic* range between the two, without knowing whether equation (22) or (23) or a combination of the two should be used to describe it. In very weak light vision is certainly scotopic, and the threshold, when the rods are thoroughly dark adapted, is for a source of sufficient extent, about 10^{-6} scotopic candela per square metre.

Besides this difficulty of visual photometry, due to retinal duality, it must be remembered that these quantities have no significance except for a retina whose spectral sensitivities are V_λ and V'_λ. It is therefore entirely meaningless, for example, to employ illumination in lux for the photobiology of an animal whose spectral sensitivity is not usually known. In certain cases it has been possible by training to determine it for higher animals. If, for example, the threshold of a dark-adapted cat is to be determined, training is begun by teaching it to open a cage above which is a light. By pressing on a pedal the cat obtains food. When the light is out and the cat presses on the pedal, it receives an electric shock. Conditioning is fairly rapid, and the cat only presses the pedal when the light is on. The light is gradually weakened and the threshold is that point below which the cat responds no more. It is found that the cat has a slightly better threshold than man, probably because its pupil opens wider, but the difference is not very marked.

The light threshold which has just been discussed is

the *absolute* threshold. There is also a *differential* threshold, in which two luminances seen, L and $L + \Delta L$ are compared, either simultaneously side by side, or in succession in the same place. The smallest value of ΔL for perception of the difference between the luminances must be determined. At a first approximation, the ratio $\Delta L/L$ is almost constant (Weber-Fechner Law) and of the order of 0·01.

The explanation of the scotopic visual efficiency by the action spectrum of rhodopsin was proposed in 1894 by König. The definition (20) of V'_λ involves energy and not number of quanta; it is therefore the quantity V'_λ/λ which defines the efficiency in photons and which should thus be compared to the action spectrum. It is true that this alters the maximum very little; it is displaced only from 507 to 504 nm. The absorption due to the media of the eye has also to be taken into account, especially the lens, which lowers V'_λ in the blue and violet. Higher values are obtained in *aphakic* subjects (with no lens), who see even ultraviolet. A small secondary maximum has been described in some of them between 335 and 365 nm; this is almost certainly related to the UV absorption band of rhodopsin (*cis* band). Another absorptive effect is that of the photoproducts of rhodopsin decomposition, the absorption of these products being in opposite directions in the retina and *in vitro*. In the retina these products form a screen and partially hinder light from reaching the rhodopsin, although in a solution of rhodopsin their absorption is added to that of the photosensitive substance (even in the difference spectra, which only partially eliminate this source of error). Finally, it must be taken into account that in the rod the chromophores are orientated so that their long dimension is at right-angles to the axis of the rod. This also effects the absorption spectrum but in a different way in the principal band and in the harmonic *cis* UV where the comparison between V'_λ/λ on aphakic eyes (measured by ERG) and the absorption spectrum of rhodopsin leads to a reduction of the *cis* band to 4/9 for the best agreement (Fig. 39).

116

All these difficulties obviously make it desirable to measure the action spectrum of rhodopsin in the rod itself. As has been indicated previously, there are two methods for this. The first consists of microscopic observation of freshly dissected retinas, and measurement of their absorption. *Micro-spectrophotometers* have been constructed which measure the spectral absorption

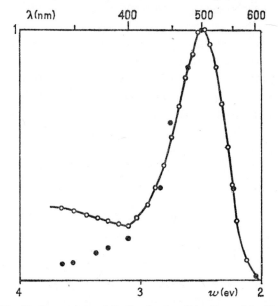

Fig. 39 Comparison of rhodopsin absorption curve (o) with scotopic efficiency V'_λ/λ (o), after Dartnall

of a surface of a few μ^2, but the intense light rapidly bleaches the rhodopsin and a complete spectrum cannot be obtained except by repeating the same operation several times on different rods. In the frog, the results agree well with the extraction methods, apart from a displacement of 2 to 3 nm towards the longer wavelengths. It can be directly confirmed that the rhodopsin is uniformly distributed throughout the outer segment, 90 per cent of the chromophores being orientated perpendicularly to the axis.

117

The second method utilizes reflectometry from the back of the eye, by ophthalmoscopic examination. Abelsdorff (1897) noted that in albino animals (with non-pigmented retina and choroid) or on those which have a *tapetum*, that is a reflecting layer behind the retina, the red colour of the dark-adapted retina, and its yellowing under the action of light from the ophthalmoscope can be observed. In most vertebrates, and in man in particular, the choroid is red and so much light is absorbed by the epithelial pigment that very little light leaves the eye. The direct method cannot be used, and the difference spectrum must be measured before and after bleaching, but the result is not very precise. Absorption by the pigment *in situ* is displaced slightly towards the longer wavelengths by comparison with rhodopsin extract. This effect is more marked when the difference spectrum is measured after a brief, intense flash, which seems to show that the displacement is due to a decomposition photoproduct which behaves differently in the eye and *in situ*.

In man, the maximum optical density of rhodopsin in the retina is about 0·15 at 500 nm, which means that at best only 3 out of 10 photons are absorbed. (In some deep-sea fish, the density is greater than 1 and the probability of absorption is increased to more than 90 per cent.) Psychophysical experiments have shown that at the threshold of vision to a light flash the entry into the eye of several tens of photons of 500 nm is required; after accounting for ocular absorption it is clear that only a small number is effectively absorbed (5 on average). Of course these few photons must be absorbed by rods sufficiently close together for their basic signals, ineffective separately, to summate by confluence on the same bipolar cell, and these absorptions should take place within an integration time of about one tenth of a second. We cannot see the individual photons, but not many are needed at the limit of vision.

Attempts have often been made to ascribe the variations in threshold during dark-adaptation to a simple accumulation of rhodopsin in the rods, which enhance the probability of absorption of photons. There are many objections to this hypothesis: for example a simple calculation shows that, even at relatively high luminances (10^{-3} candela per square metre for instance), each rod waits several seconds between the absorption of two successive photons, and, as there are millions of molecules of rhodopsin per rod, when the retina is thoroughly dark adapted, moderate light intensities are practically unable to vary the rhodopsin content of the retina. Now the threshold varies very obviously under these conditions: a luminous background so weak that during the experiment 1 rod in 10 at the most can absorb a photon, is sufficient to increase the absolute threshold by a factor of 3. The rise in threshold is not therefore due to a change in the rods themselves, but to a modified interaction which necessitates a greater number of absorptions to reach threshold. It is clear that the concept of quantum efficiency loses much of its significance in experiments such as these. It is therefore a modification of the nervous organization which is related to variations in threshold, this modification being either a signal from the bleached rhodopsin, or an effect of the change in the rate of regeneration of rhodopsin as a function of time. In the whole field of scotopic vision this nervous mechanism would in fact be the only control, as the rhodopsin level in the rods does not begin to drop until photopic levels are reached, or until it is predominantly the cones which are in action.

The interpretation of the photopic curve V_λ is much less direct; it will be discussed later in relation to colour vision. It has been possible to measure V_λ in the infrared up to 1,050 nm ($V_\lambda = 10^{-13}$ approximately) and it is remarkable that, after correction for absorption due to the media of the eye, which becomes important in this region because of the absorption bands of water (at 1μ only half the radiation which penetrates the eye reaches the retina), the photopic efficiency at long wavelengths obeys a very simple law. The logarithm of V_λ/λ is a linear function of w (or of the frequency). It is equally surprising

119

that the scotopic function V'_λ obeys the same law in the long wavelengths, the straight lines even being parallel. It is probable that this fact is due to the absorption of photons of weak w acting on the chromophores of pigments by a mechanism which sets in motion molecular rotations and vibrations; an interesting consequence is that temperature must affect the threshold a little in the long wavelengths and in fact a hot bath which raises the temperature in the eye by $2°$ improves sensitivity in the extreme red.

If the infrared can be made visible by augmenting the energy flux, it is known that the UV down to $0.3\ \mu$ acts on the retina if the lens is removed. Similarly X-rays are weakly seen. Consequently the concept of limits to the visible spectrum becomes imprecise: the classical values (0.4 to $0.7\ \mu$) simply indicate that V_λ has fallen below 0.001.

One question which has caused a good deal of confusion is that of the biological adaptation of V_λ and V'_λ to the spectral distribution of the environment. It has already been frequently pointed out that the position of the maximum of V_λ and V'_λ (or rather the quotient of these quantities divided by λ, which is little different) has a real significance, whatever scale is used as abscissa. On the other hand, for the spectral distribution of energy the maximum is deprived of meaning (in particular there are four different values of λ_m according to whether the abscissae are measured in λ or in w, and the ordinates in energy densities or in number of photons), and the fortuitous coincidence of the maximum of V_λ with the sun's energy at earth level does not make biological sense. The most logical photochemical schema (N_w as a function of w), places the maximum energy of the sun's radiation in the infrared, but a V_λ maximum in this region would not be expected because of marked absorption by water, which is the essential constituent of the transparent media of the eye.

For marine animals, the problem appears simpler at first sight. I have calculated the sun's energy (Fig. 40) in photons reaching a depth of 5 m (solid line curve) and 50 m (dotted curve) in clear coastal water (Jerlov's type I). It is tempting to say that a pigment of the rhodopsin type is perfectly adapted to life at a depth of 50 m, and that when the fish rises towards the surface a Purkinje effect would explain how, while the light is increased, the position of the N_w maximum is displaced from 500 to 560 nm. But this would be to fall into the same error as

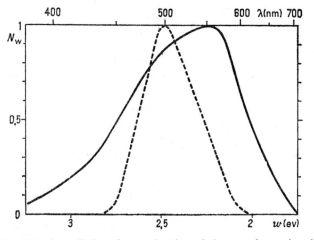

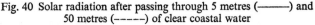

Fig. 40 Solar radiation after passing through 5 metres (————) and 50 metres (-----) of clear coastal water

before. The logical method consists in seeking the maximum of the quantity:

$$\int N_w \times \alpha_\lambda \times dw \tag{24}$$

denoting by α_λ the coefficient of extinction of a visual pigment (assumed to be diluted) which is supposed to belong to the family defined by Dartnall's nomogram. This calculation of variations has an absolute meaning, independent of any arbitrary convention. The most favourable position of the pigment would be close to $\lambda_m = 545$ nm at a depth of 5 m and therefore confirms the previous interpretation of the Purkinje phenomenon.

121

It is worth noting that equation (24) defines a 'biological' number of photons and there is every chance that these numbers are additive when several distributions of N_w are superimposed: this would be the rational form of Abney's law, which would thus find its theoretical justification, at least when there is only one retinal pigment.

6.3. Visual acuity

The recognition of shapes constitutes the principal function of visual activity, at least in animals with eyes. Only a few brief comments on this important subject will be made.

The anatomical structure of eyes obviously constitutes the basis of this study. The principal types are as follows: the compound eye of the insect (Fig. 41a) is made up of separate ommatidia, each one localizing a portion of the visual field situated round about its axis; in molluscs the formation of a rudimentary image on the principle of the 'camera obscura' appears in *Nautilus* (Fig. 41b), but optical arrangements giving a real image are also found, as in the cephalopod eye (Fig. 41c) which is strangely reminiscent of the vertebrate eye (Fig. 41d). One curiosity: the marine copepod *Copilia quadrata*, which lives in the plankton in the Bay of Naples; the female has eyes which occupy most of its body, and each one comprises a lens (Fig. 41e) giving an image in the plane of which a single rhabdom moved by muscle action scans the image, on a principle which is the forerunner of television.

Behaviour experiments made on insects with a visual target made of parallel black and white lines have been interpreted as a proof that *visual acuity*, defined by the angle between two consecutive lines which just permits their resolution, is related to the angle between two neighbouring ommatidia (of the order of a degree). This mosaic theory is not supported by electrophysiology. In recording the responses of isolated ommatidia, it has been shown that the fields of neighbouring ommatidia overlap considerably. It is even possible that in an ommatidium the visual retinula cells function independently and may be able to analyse a little the rough image given by the dioptric

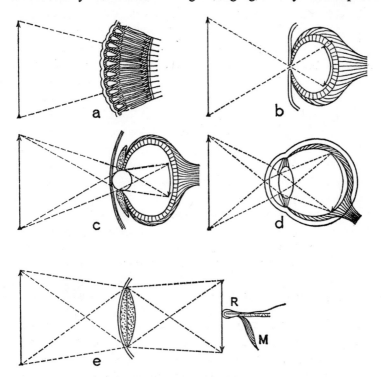

Fig. 41 Various types of eye (diagrammatic)
(a) insect compound eye; (b) 'camera obscura' (*Nautilus*);
(c) cephalopod; (d) vertebrate; (e) *Copilia quadrata*
(the muscle M moves the single rhabdom R in the plane of the image)

system of the ommatidium. When a light is shifted, some visual cells in the ommatidium 'see' at any given moment, others do not. The state of knowledge on this question of how the insect sees shapes is still very little. Probably recognition of detail is mediocre, this deficiency being compensated for by a keenness,

123

superior to that in vertebrates, in the perception of rapid visual changes. If a light is interrupted many times per second, some insects still react to the flicker when the frequency becomes greater than 200 per second from electrophysiological recordings, although in man the *critical fusion frequency* stops at considerably less than 100 (luckily for the cinema and television).

Electrophysiology makes important contributions to the study of the visual acuity of invertebrates. In the well-known

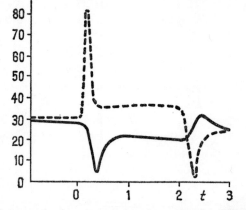

Fig. 42 Interaction between two *Limulus* ommatidia, after Ratliff

crab *Limulus*, inhibition between neighbouring ommatidia has been confirmed, which enhances apparent contrast and visual acuity. These effects are quite complex, as can be seen on the record (Fig. 42) of two adjacent receptors. In one the response, measured by the number of spikes per second, is marked by a solid line, this receptor receiving a certain illumination E, constant during the recording. The other receptor (dotted line), also receives illumination E at first; at zero this illumination is abruptly increased for a duration of 2 seconds, and is then returned to illumination E. The records show simultaneous and remarkable variations whose result is clearly to increase the apparent difference between the illumination of contiguous receptors, by a *simultaneous contrast* effect.

124

In vertebrates visual acuity is a function of the optical quality of the image given by the ocular system on one hand, and the anatomical and physiological capabilities of the retina in analysing this image on the other. The human retina, for example, contains about 7 million cones and 120 million rods, but there are not even 1 million fibres in the optic nerve to conduct the information, and it seems that each fibre can transmit only one type of response, through the variation in frequency of the impulses (spikes) which are propagated in it and which are identical with each other. It is worth noting in passing that nature has utilized for transmission the principle of frequency modulation, and not of amplitude modulation, for the same reason that it is chosen in modern radio technique, that is, the suppression of parasite signals. This relative poverty in the transmission capacity of the optic nerve has been compensated for in two ways: the first is the existence of a favoured region in the retina (fovea) where the cones are densely packed and have in addition their own transmission path to the brain. In man it seems as though the density of the foveal cones would be sufficient to take advantage of the physical quality of the retinal image, which in the most efficient eyes is only limited by the wave structure of light. Some birds even have two foveas on each retina, one permitting binocular vision in front, the other lateral monocular vision (Rochon-Duvigneaud's 'visual trident' in birds of prey).

The second method consists of utilizing the movements of the eyes. In man the eyeballs make continual rapid movements (about 100 per second), of small amplitude (1 minute of arc) which sweep the details of the image over several contiguous cones. If the retinal image is stabilized by an optical artifice, the contrasts rapidly become less evident and the image, not being renovated, disappears into a uniform grey.

Electrophysiological studies on vertebrates have shown important nervous interactions which, more than in *Limulus*, act to refine the image, by enhancing contrasts and aiding perception of changes. In the cat there are receptive fields in the retina which have either *on* type or *off* type receptors at their centre, and the reverse at the periphery. In dark adaptation, these distinctions disappear, which proves yet once more that the retinal connections are reorganized in relation to luminosity. In summary, our daytime retina (cones) is mainly organized for the discernment of detail; our night-time retina (rods) for vision in weak light, not only due to the choice of different receptors which, for scotopic vision, seem richer in pigment and are much more numerous, but also by a different system of transmission of information by the visual cells, the cones being 'individualists', which mutually inhibit one another, the rods more 'sociable', uniting their effort to see all together.

Although it is outside our scope, it may perhaps be interesting to the reader to know that a certain structure of the receptors, notably related to the vertical and horizontal directions which play a fundamental role in animal vision, can be followed as far as the brain. Certain receptive areas are specially sensitive to stimuli with vertical symmetry, and others to those with horizontal symmetry. These phenomena are even more marked in invertebrates. The recognition of simple shapes has been extensively studied in the octopus, and the tendency of this creature to order the universe according to the vertical or the horizontal is extremely clear. Visual and tactile memory mechanisms in the octopus are the best (or perhaps the least poorly) known, and vertical and horizontal memorization seems particularly important in this animal.

REFERENCES

FUORTES, M. G. F., J. Physiol. **148**, 14 (1959)

JERLOV, N. G., Optics of Sea Water, in *International Dictionary of Geophysics*, Oxford, Pergamon (1968)

KENNEDY, D., J. Gen. Physiol. **44**, 277 (1960)

LE GRAND, Y., *Light, Colour and Vision*, 2nd ed., London, Chapman and Hall (1968)

LE GRAND, Y., *Form and Space Vision*, Bloomington, Indiana University Press (1968)

NAKA, K. and EGUCHI, E., J. Gen. Physiol. **45**, 663 (1962)

RATLIFF, F., HARTLINE, H. K., and MILLER, W. H., J. Opt. Soc. Amer. **53**, 110 (1963)

ROCHON-DUVIGNEAUD, A., *Les yeux et la vision des vertébrés*, Paris, Masson (1943).

SVAETICHIN, G., Acta Physiol. Scand. **39**, suppl. 134 (1956)

TOMITA, T., Japan J. Physiol. **6**, 327 (1956)

CHAPTER VII

Colour Vision

7.1. Human and animal colorimetry

By a fortunate chance, man is able to see colours: the spread of the visible radiation spectrum is appreciated as a succession in which continuous qualitative changes from violet to red can be distinguished. (Table I, beginning of Chapter I.) If he had lacked this facility he would probably not have investigated it in other species, and the present chapter would not have been written. Electrophysiological results, which we interpret rightly or wrongly as proof of colour vision in various animals, would then have appeared as strange anomalies.

This is, however, not so, and men have been interested by colour throughout the ages, always using it for artistic purposes, from prehistoric cave paintings to the polychrome statuary of classical Greece. The ancients thought of colour as a mixture of light and darkness. This is true in the sense that a red object, for example, is light in the long wavelengths, which it reflects, and dark in the short, which it absorbs. On the other hand the dyeing industry has found by experience that almost any hue can be reproduced by a mixture of the 'primary' colours (red, blue and yellow). These three colours have been thought of as the basis of light, and followers of Newton's emission theory postulated three types of luminous particles.

The brilliant Thomas Young was the first to state that trichromatism was retinal (1801), therefore physiological and not physical. Young was a doctor, who wrote a thesis on audition, but he was also a physicist, and had made the first experiment on interference ('Young's slits'), which convinced him of the undulatory wave nature of light. Light waves, like sound waves, form a continuous spectrum. It is inconceivable, wrote Young, that each point on the retina can contain 'resonators', capable of vibrating in unison for each wavelength of visible light; it is possible for the ear, since it does not localize its sources by the projection of images on to a receptor (acoustic localization, which is very rudimentary, utilizes only the difference between the sounds arriving at the two ears); the ear is capable of delicate spectral analysis and, in a chord, the musician can recognize the component notes. The eye cannot do this: as the image must be spatially analysed the eye cannot also transmit the frequency. The eye cannot thus analyse the frequency of the photons since the information capacity of the optic nerve is inadequate for both transmissions. Young concludes that it must therefore be supposed that each point on the retina contains three receptors, one sensitive to long wavelengths in the visible spectrum, the second to medium ones, and the third to short.

For fifty years, Young's concept of trichromatism remained neglected and not understood. It was left to the researches of Maxwell, Helmholtz and Grassmann in the mid-nineteenth century to vindicate it. Helmholtz explained in particular the difference between the subtractive colour mixtures of painters, and the additive light mixtures of physicists (and of *pointillistes* such as Seurat); he showed that the *fundamentals*, that is, those colours which excite only one of the retinal receptors, must be red, green and blue. This laid the basis of colorimetry.

Sttlricy speaking a distinction should be drawn between trichromatism and the trichromatic theory. Trichromatism is accurate and certain; it states that with any four lights visual identity can *always* be produced, either by mixing three of them in suitable proportions and comparing this sum to the fourth, or by equalizing the sum of two of the lights with the sum of the other two. Mathematically this experimental fact expresses the trivariant nature of the visual system in man. The trichromatic theory is only one of the possible explanations of this fact, the most direct and probably the most likely.

It is true that colour vision in man varies from subject to subject, depending in particular on the amount of yellow macular pigment; age also affects it by yellowing the lens. These variations can be evaluated by *Rayleigh's Equation* which consists in making a yellow light and a mixture of red and green light appear the same to the subject; some subjects show marked deviation from the average, in using either too much red, or too much green. They are known as *abnormal trichromats* and they are classed as *protanomalous* and *deuteranomalous* observers. An anomaly more distinct from colour vision is known either as *Daltonism*, after the chemist Dalton who described his own case at the end of the eighteenth century, or as *dichromatism*. The colour of the universe is bivariant to dichromats, and for them all possible colours exist in the spectrum of monochromatic radiation. From the descriptions of the rather rare subjects who have one normal eye and one dichromatic, it appears that they see with their abnormal eye a two-coloured spectrum, yellow for the long wavelengths and blue for the short, separated by a colourless zone in the green. In its most typical form, dichromatism is a reduction of normal trichromatism, through the loss of a fundamental or the fusion of two fundamentals into one. Anomalies of colour vision are congenital (anomalous acquired cases are also described) and hereditary. They almost always affect male subjects, with transmission from the grandfather to the sons of his daughters. Women only show the defect if it exists in the families of both their father and mother.

Colorimetry, a technique of considerable practical importance in a world where colour plays a growing role, will not be discussed. There are many colorimetric systems, which all represent normal colour vision equally well. Grassmann showed that all colorimetric laws obey (at a first approximation) linear relations. Consequently if by carrying out experiments where lights of different colours are mixed, three wavelength functions,

$R(\lambda)$, $G(\lambda)$, and $B(\lambda)$, known as primaries, are established, the sum of which represent the resultant mixtures, then all the functions obtained by linear combinations of these three functions can equally well constitute primaries. The problem of obtaining the fundamentals, that is, the sensitivities of Young's three retinal receptors as a function of λ, is thus beset by triple mathematical indeterminacy.

An attempt to overcome this has been made by postulating additional hypotheses. The first is *Abney's Law*, according to which the luminances are additive (this is approximately true), which in effect means that the photopic luminous efficiency is also a linear function of the primaries:

$$V_\lambda = \rho R(\lambda) + \gamma G(\lambda) + \beta B(\lambda) \qquad (25)$$

The second makes use of colour vision in dichromats. In this group, *protanopes* have a reduced spectrum in the long wavelengths and they are scarcely able to see red traffic lights. They thus lack the 'red' fundamental, sensitive to long wavelengths. Another variety of dichromat, *deuteranopes*, have an almost normal V_λ, and they either lack the normal 'green' fundamental (König's hypothesis), or they possess the red and green fundamentals but these function together, which diminishes by one the variance of their colour vision (Fick's hypothesis, 1879). Finally there is a third very rare type of dichromat, *tritanopes*, in whom the 'blue' fundamental is lacking.

This analysis leads to the following results. The blue fundamental is represented by a bell-shaped curve which although narrower resembles Dartnall's nomogram curve, with a classical maximum λ_m of about 440 nm. The factor β in equation (25) is very small, which indicates that this blue fundamental, which is essential for colour vision, plays only an insignificant part in luminance. With König's hypothesis, the green and red fundamentals have their classical maximum λ_m at about 540 and 560 nm respectively and the red fundamental curve differs distinctly from a normal pigment curve. In addition, $\rho = 0.87$

131

and $\gamma = 0.13$, which indicates that the red fundamental plays a predominant role in luminance. With Fick's hypothesis, on the contrary, the λ_m maxima are about 540 and 580 nm respectively, the curves agree better with Dartnall's nomogram and the participation of the red and green fundamentals in luminance is better balanced ($\rho = 0.44$, $\gamma = 0.56$). It is necessary to realize that these arguments are rather tenuous and that firm conclusions about the fundamentals of the human retina cannot be drawn from them.

It was hoped that Stiles' *bicolour threshold* method would be of value in deducing the fundamentals from subjective experiments in man. On to a uniform field of wavelength λ and of fixed energy value, a light of wavelength μ, which occupies only a small surface of the total field, is projected. The subject determines at what energy level of radiation μ threshold is reached. If $\lambda = \mu$ it can be seen that an ordinary differential luminance threshold is involved, although if $\lambda \neq \mu$ a complex function is being studied. A large number of careful experiments for all combinations of λ and μ gives curves which show breaks interpreted by Stiles as indicating the transition from one mechanism to another in colour vision, in the way that the break in Figure 37 (p. 111) shows the transition from cones to rods. There would in effect be three types of colour mechanism with λ_m of about 440, 540 and 580 nm, which is in reasonable agreement with the Fick-type fundamentals. Each of these mechanisms, however, takes two or three forms of spectral sensitivity according to the conditions of adaptation, and in addition if an attempt is made to calculate the colorimetry of the subject using Stiles' fundamentals, the results are clearly abnormal. Stiles' mechanisms, in spite of their great interest, cannot correspond to the fundamentals. The phenomenon of the bicolour threshold results from a complex series of interactions and inhibitions from which photochemical simplicity has disappeared.

In animals, colorimetric analysis is difficult and unreliable.

The phenomena of homochromia have often been interpreted as a proof of colour vision. For example the shrimp *Hippolyte varians* seeks shelter under seaweed the same colour as its body, but this is a defence reaction utilizing a radiation filter, which does not prove anything about its perception of chromatic qualities. The modification of body colour to harmonize with the background (chameleon) is likewise unconvincing. First it must be proved that it is not what Piéron calls *homoleucia*, that is, a harmonization of reflection factors resulting by chance in the similarity of colours. The ocular basis of homochromia would then have to be demonstrated.

Phototaxis in planaria and daphnias has also sometimes been interpreted as a sign of colour discrimination, but Viaud has demonstrated that it is simply a question of a variation in the luminosity of the radiation. The spontaneous discrimination of some animals in alimentary or reproductive reactions does not prove anything either. The male cabbage butterfly reacts to yellow by coupling reactions, at the mating period, and the female, at the time of egg-laying, reacts to green surfaces; it is quite possible that this correlates simply with the light appearance (of yellow) and the darker one (of green) of the natural pigments to this creature.

A more elaborate method makes use of *optokinetic nystagmus* (von Buddenbrock, 1927): the animal is placed in the middle of a slowly turning drum which has coloured vertical bands on it, and any reaction to their movement is looked for. For example the animal reacts strongly to black and white bands. It reacts less to light and dark grey bands, and not at all to bands of nearly the same grey (a differential luminance threshold can be determined in the same way) since it sees the cylinder as uniform and perceives no movement. With yellow and grey bands, if a reaction occurs it is because the yellow is perceived by a quality of its own, different from the range of the greys; the same is true for blue and grey bands. In fact there will be a minimum reaction for yellow and one particular tone of grey. If a minimum is also observed for the same grey and a particular shade of blue, then an increased reaction of the animal to the yellow and blue bands which have the same luminosity and differ only in colour would indicate the appreciation of colour. It can be seen that these experiments need patience and care to be really convincing.

In the higher animals, training is possible, similar to that previously outlined in connection with the determination of the luminous threshold. But here also caution is needed. It is easy, for example, to train a dog to run after a red ball and not after a blue ball, but this is because it sees the first as dark grey and the second as light grey. In order to be convincing it would be necessary to train it to recognize the red ball among the whole range of grey balls from light to dark. This does not seem possible, and one concludes that apparently the dog does not see colours. While this method can be used to demonstrate that an animal possesses some chromatic discrimination, it can never be definitely proved that it lacks the power entirely.

In the bee, von Frisch and his collaborators have been able to demonstrate conclusively the presence of colour vision. A chequered board composed of squares of different greys with one blue square among them is used and a cup containing a sugar solution which attracts the bee (Fig. 43) is placed on the blue square. The arrangement of squares on the board is changed daily, which trains the bee to recognize the blue among the range of greys. Then the cup is removed, and a large pane of glass placed over the board, to eliminate all but visual clues. The bee still comes to the blue square, wherever it is positioned which demonstrates that it can see blue. On the other hand, it cannot see red, which it confuses with dark grey. The existence of colour contrast can even be demonstrated. Trained to the blue, it will come to a grey square on a yellow background, the blue being induced on the grey by contrast with yellow.

Which animals show colour vision? In spite of homochromia, nothing can be said about some of the cephalopods. In some

decapod crustaceans, chromatic vision is probable (optokinetic nystagmus). It is definitely present in teleost fishes
(minnow, gudgeon, carp, etc.) which show clear analogies with
human colour vision. It is also present in some bactracians

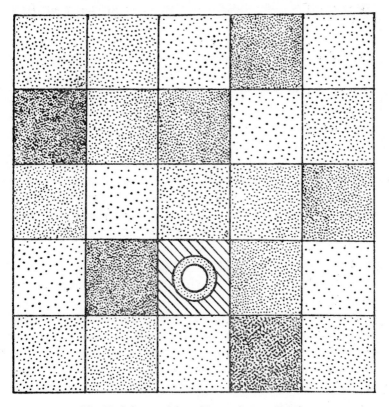

Fig. 43 Colour training of bees, after von Frisch

(frog), in some reptiles (tortoises and lizards) and in diurnal
birds. Mammals with colour vision are rare, and only some
tree-dwelling and fruit-eating species seem to be supplied with
chromatic discrimination (squirrel, primates): biologically this
is of use to them in their choice of food, a problem which does

135

not present itself to herbivores and carnivores. Perhaps the horse has the rudiments of colour vision, and also the dog(?). The irritation of the bull by red is a fable, it will charge just as well at a white *muleta*. In short, colour vision is a luxury reserved for a small number of vertebrate species, apart from birds.

On the other hand it appears that many insects have chromatic discrimination, but somewhat different from that of vertebrates. In 33 different genuses, electrophysiology, the optokinetic reflex and training have been used to prove that colour vision exists. In general the visible spectrum is displaced about 100 nm towards the short wavelengths compared with that of vertebrates. Insects see ultraviolet, not usually red.

Colour vision has been studied especially in the bee, which can be trained, as explained above. Daumer has been able to show that the bee has trichromatic vision; for example radiation at 360 nm and radiation at 490 nm are complementary, their mixture in suitable proportions giving white (defined by the spectrum of the xenon arc); bees easily recognize three regions in their spectrum (300–400, 400–480, 500–600) and a narrow transition band 480–500 which plays a similar role to our yellow.

In the fly *Calliphora*, it appears that the spectrum is extended further into the red, nearly to 700 nm, but 'protanope' flies have also been observed which do not see red. Besides, in many insects colour vision is not uniformly distributed. Sometimes the creature sees the colour low down, but not in the top of the visual field.

Colour vision in insects plays an important part in the pollination of flowers; in particular, petals which seem white to us are seen as strongly variegated by the bee, which can guide itself on sap (which absorbs UV) patterns to plunder the pollen. As for the problem of knowing whether retinal duality (photopic and scotopic pigments) exists in insects or not, it has not been solved, but specialists incline to the belief that it does not.

7.2. Pigments and colour vision

The simplest interpretation of Young's trichromatic theory consists in supposing that there are three types of retinal cones, each type having its outer segment impregnated with a different pigment. The objection might be made that the cones are all anatomically the same, or at least that their form only varies with distance from the fovea, and it might be thought that a variation in pigment would be accompanied by a variation of form (in this way the rods of the frog are of two types, as regards both form and pigment). Obviously it might be supposed that each cone contains the three pigments, but that they act differently (chemically or electrically) on the bipolar cells. Another objection is that pigments have never been definitely extracted from cones.

Various hypotheses have been advanced to explain these difficulties. For example, there might be only one cone pigment, which might react differently to light of different wavelengths. The cones might constitute *dielectric wave guides* for radiation, with various possible modes of resonance; actually a wave guide phenomenon, the 'Stiles-Crawford effect', concentrates light in the axis of cones. But when light reaches the cones in some unusual way (across the sclerotic or after diffusion from the blind spot), its apparent colour does not change; Murray (1968) has also studied a homogenate of foveal outer segments of Rhesus monkeys and he found that it contained more than one pigment.

In various animals (tortoise, chicken, pigeon), there are coloured fatty droplets between the inner and outer segments of certain cones. They have sometimes been considered as filters which would permit colour discrimination, as in the autochrome plates made for colour photography by Lumière in 1906. In the chicken, particularly, carotenoids of three different types occur. It is possible that these droplets interfere with colour vision by enhancing colour contrasts in some parts of

137

the visual field, but it seems fairly improbable that they are by themselves sufficient. I have in fact calculated what sort of colour vision the chicken would have according to this theory. It would be almost dichromat, which is contrary to behaviour observations. On the other hand the droplets vary with adaptation, and their colour is also modified, which indicates that the globules might perhaps be reserves of products used in synthesizing the cone pigments.

The method of microscopic examination of retinas freshly removed from the eye has recently yielded interesting results. By analysing in a microspectrophotometer circular areas of 0·2 mm in diameter in the retina of the Rhesus monkey and man, differential spectra are obtained from the fovea which suggest the existence of 'green' and 'red' pigments, the second in amounts about twice those of the first; the density variation after bleaching remains small (of the order of 0·04, or 4 times less than for rhodopsin).

The first microspectrophotometer sensitive enough to determine the spectral absorption of isolated cones, without excessive distortion due to bleaching during measurement, was constructed by Marks and tried out on the goldfish (1963). This creature has large cones and it is also known from behaviour experiments that it sees colours. The position of the classic λ_m of 113 cones of the goldfish is indicated in Figure 44: there are three obvious difference spectra accumulations at about 455, 530 and 635 nm; some intermediate values are probably due to two neighbouring cones being measured at the same time. The three absorption curves are similar in form, which makes it unlikely that the central group, which is numerically the most important, contains, in addition to 'green' cones, composite receptors where the three pigments are mixed. Young's theory thus, after a gap of more than 160 years, has its first real experimental confirmation, and at the same time it is confirmed that the three pigments are contained in different cones.

In the monkey and man similar measurements have been

attempted, but the small size of the cones (especially in the fovea) makes such measurements uncertain; it appears, however, that three types of cone do exist, with λ_m near those anticipated in colorimetry by Fick's hypothesis. Improvement of the technique will doubtless provide a definitive answer in the future.

Fundus reflectometry has been used to examine the human fovea, but the technique is difficult, in part because of diffused light. Weale uses an apparatus in which the incident light from

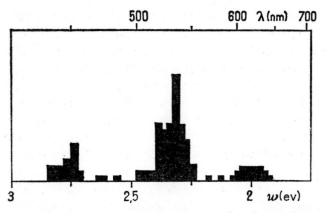

Fig. 44 Absorption by the cones of the goldfish, after Marks

a xenon arc passes successively through 26 interference filters ranging from 405 to 680 nm, the complete cycle lasting less than 0·3 second. This light is concentrated on a surface of about 1° diameter, centred on the fovea. The light which leaves the eye is directed on to a photomultiplier connected to an oscillograph. Another optical device sends light, which causes bleaching, into the eye on a field of 3° which overlaps the field being studied. The complete measurement of the difference spectrum lasts 3 seconds. However, it is difficult to avoid eye movements in the subject, in spite of the fixation point imposed.

As has been already stated, the absorption due to the photoproducts of pigment decomposition makes the interpretation of

139

results difficult. In the case of the grey squirrel, an animal which has only cones, reflectometry gives a difference spectrum whose λ_m is at 535 nm. The density variation due to bleaching is between 0·08 and 0·16 according to whether the light is considered to cross the cones twice or not and this is uncertain because of the intervals between the cones. This animal is the only one from which it has been possible to extract a cone pigment. This has a λ_m at 504 nm although the photoproduct maximum is at about 480 nm, the difference in these curves giving an apparent maximum between 530 and 540 nm. The density variation of the extracted pigment would correspond, in the retina, to a change of only 0·05 or 15 or 30 times less than the result given by reflectometry. It is possible that the marked difference is due to the concentration of light towards the point of the outer segment of the cones by the wave guide effect (or, if the language of optical geometry is preferred, by total reflection within the cone) and to the accumulation of pigment towards the apex of the cone.

In spite of these difficulties, reflectometry from the fundus of the eye has made possible, for the first time, the demonstration, by Rushton, of the existence in the human fovea of at least two photosensitive pigments. He called them 'chlorolabe' and 'erythrolabe', meaning absorbing green and red respectively. The first is appropriately named since the λ_m is about 540 nm, although for the second pigment (580–590 nm) the maximum falls more in the orange-yellow. The 'chlorolabe' pigment is the only foveal pigment in protanopes (red-blind dichromats); in other dichromats named deuteranopes the situation is not known with certainty. It has not been possible to detect the third pigment (λ_m in the blue) by reflectometry, but the fovea probably contains little, and in addition functions almost as a dichromat of the tritanope type, which raises the possibility of rhodopsin being the pigment of the 'blue' receptors, but this hypothesis seems unlikely.

Although colour vision necessitates the presence of at least two pigments, the demonstration in an animal retina of two different pigments does not imply that it sees colours. It simply indicates (in general) day-retina/night-retina duality. Obviously Purkinje's phenomenon permits, in a certain sense, chromatic discrimination: a total dyschromatope, that is, a subject whose cones do not see colours, may, in equalizing a blue and a red with photopic vision, subsequently confirm with scotopic vision that the blue is lighter and the red darker, and deduce that he is looking at blue and red, but it is not a question here of properly speaking colour vision (in the same way the Daltonian electrician can use coloured glass to help him recognize the colours of the wires in a circuit).

If everyone agrees in attributing dichromatism to a deficiency in the trichromatic system (the absence of a pigment, or the confusion of the response of two pigments), the causes of abnormal trichromatism are less clear. It has sometimes been supposed that it might be a question of an excessive or insufficient concentration of a normal pigment, but it is easy to show that this has no effect on the colorimetry of the subject. Besides, an intense colour adaptation, to red for example, bleaches the 'red' pigment more than the others, and because of this everything appears greenish, but colour equations (Rayleigh's for example) are unchanged unless the adaptation is so intense that very little pigment and a large amount of photoproduct remain. Therefore a modification in the pigments themselves must necessarily be postulated. There seem besides to be dichromatic subjects who are at the same time abnormal: they no longer accept the colour equations made by normal trichromats, but those of a type of abnormal trichromat. All these defects are hereditary, and at present incurable.

7.3. Chromatic electrophysiology

Electrical visual recordings all depend on wavelength, but obviously this does not indicate that they show colour vision.

For there to be chromatic discrimination, the impossibility of an equilibrated substitution must be shown. Consider monochromatic radiations of two wavelengths, λ and λ': if a single photosensitive pigment exists, in the visual receptor whose electrical response is being measured, this receptor reacts identically to the photons which it absorbs, and these only differ in their probability of being absorbed. It is therefore possible to compensate for this probability by sending unequal numbers of photons N and N' (in a given time) on to the receptor, and the response will be identical as long as the absorption curve, which is a measure of this probability, does not vary (assuming that the pigment is dilute enough). This substitution of wavelength λ' for wavelength λ still has no effect on the response, whatever the pigment concentration, and therefore, possible preliminary bleachings. If this substitution does not remain in equilibrium as a function of adaptation, it is because there are two different pigments, a necessary (but not sufficient) condition for chromatic discrimination.

In the fly *Calliphora*, a similar method allows a presumed colour vision to be confirmed. In alternating λ and λ' it is impossible, whatever the value of N and N', to eliminate fluctuation of the total electrical response (ERG). On the other hand, using microelectrodes, three maxima can be established, at about 340, 490 and 553 nm, but this does not prove trichromatism. It is possible that 340 and 533 could be two maxima of the same pigment. All that can be said is that the fly is at least dichromatic.

In the bee, which is known by training to have trichromatic vision, and from which a pigment ($\lambda_m = 440$ nm) has been extracted, electrophysiological studies show two receptors in the compound eye of worker bees (340 and 535 nm), this time separable by selective adaptation. This is the only invertebrate in which there is firm evidence for trichromatism.

A large amount of research has been done on vertebrates. In particular, Granit and his collaborators have recorded volleys of spikes from the ganglion cells as a function of λ and describe two types of response: those of the *dominators*, the most frequent, are the curves which resemble in form the luminous efficiencies V_λ and V'_λ in man; these are photopic or scotopic luminosity receptors. Probably there is no cone for luminosity, and it is simply a question of the confluence on one bipolar cell of the responses of three types of cones when there is colour vision. A long-standing objection to this interpretation was that the addition could be explained at the level of one cone if it contained a mixture of pigments, but that it was difficult to imagine a linear addition of the nervous responses. This argument is no longer tenable since the ERP of visual cells where linearity is conserved was discovered. Obviously if the animal does not have colour vision, its dominator reflects simply the absorption by the pigment in its cones.

The other form of response is known as *modulator*. It produces pointed curves, with maxima in diverse spectral regions. This is probably another complex response, but, in contrast to that of the photopic dominators which can result from an addition of responses, in this case a differential effect between the fundamentals occurs. The maxima of the modulators, although varied, are grouped in three main regions. In some fish, the modulators are displaced towards the red (compared with those of the frog) by a constant quantity of the w scale, which is due to replacement of retinene$_1$ by retinene$_2$. In the pigeon, a displacement towards the red could also be due to coloured droplets.

With micro-electrodes inserted into the cones of the carp, Tomita *et al.* (1967) recorded potentials which appear to represent the electrical response of the inner segments of the photoreceptor; they show, as a function of the wavelength, a spectral variation where the three different types of cone of Young's theory are again found. In certain fish, recording of the S potentials gives three different curves as function of wavelength (Fig. 45): a receptor of luminance L, and two receptors of chromaticity C with antagonistic responses, one giving a positive potential for the long wavelengths, and negative in the centre of the spectrum (red-green pair RG), the other positive in the blue and negative in the yellow (BY pair). This mechanism recalls the old colour theory of Hering which formerly seemed exclusive of Young's theory. This is not so, since Young's trichromatism governs what occurs at the level of the cones while Hering's antagonistic pairs operate farther on in the

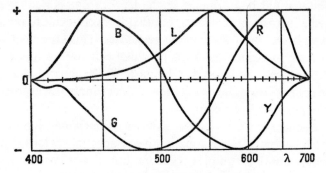

Fig. 45 S potentials in fish, after Svaetichin

nervous coding of colour. It is not known whether this is so in the retina of man and the monkey, but a similar mechanism is present at the level of the *lateral geniculate nucleus* (LGN) which is the relay between the eye and the cortex. Monochromatic stimuli falling on the monkey's retina evoke two types of response in the LGN cells: some cells have a wide range of spectral response (either inhibition of spontaneous activity in the dark, or excitation, which increases the spike frequency), others have antagonistic responses, some radiation augmenting and other diminishing the frequency. When the light intensity is altered, the reaction differs according to whether inhibition or excitation is in question, which probably explains the so-called Bezold-Brucke's phenomenon, that is, a systematic effect of the intensity on the apparent hue of monochromatic radiation. Chromatic adaptation of the retina acts

144

very little on the cells with a wide response, and considerably on those with antagonistic responses, which agrees with well-known subjective facts.

Finally, colour vision, which arises in the cones due to the presence in them of three different pigments, is transformed in the retina itself into one impulse for luminosity and various impulses for colour and elaborated by confluence of the impulses transmitted by the bipolar cells on the ganglion cells. These colour impulses undergo 'recording' in the LGN so as to be better assimilated by the visual cortex; possibly this coding makes use of the spontaneous frequency of the LGN neurons as a kind of carrier frequency modulated by the signal, the decrease and increase of this frequency carrying complementary information, conforming to the subjective aspect of colour.

REFERENCES

DAUMER, K., Experientia **16**, 289 (1960)
VON FRISCH, K., *Bees, their vision, chemical senses and language*, Ithaca, Cornell University Press (1950)
GALIFRET, Y., ed. *Mechanisms of Colour Discrimination*, Oxford, Pergamon (1960)
GRANIT, R., *Receptors and sensory perception*, Newhaven, Yale University Press (1955)
LE GRAND, Y., Vision Res. **2**, 81 (1962)
MARKS, W. B., Ph.D. Thesis, Baltimore, Johns Hopkins University (1963)
PIERON, H., *Psychologie zoologique*, Paris, Presses Universitaires (1941)
MURRAY, G. C., Ph.D.Thesis, Baltimore, Johns Hopkins University (1968) *Soc. Adv. Sci.* Washington, 803 (1959)
PIRENNE, M. H., *Vision and the eye*, 2nd ed., London, Chapman and Hall (1967)
RUSHTON, W. A. H., J. Physiol. **168**, 345, 360, 374 (1963); **176**, 24, 38, 46 (1965)
TOMITA, T., Vision Res. **7**, 519 (1967)
WEALE, R. A., Doc. Ophthalm. **19**, 252 (1965)
WRIGHT, W. D., *Researches on normal and defective colour vision*, London, Kimpton (1946)
WYSZECKI, G., and STILES, W. S., *Color Science*, N.Y., Wiley (1967)

CHAPTER VIII

Non-visual Reactions to Radiation

8.1. Direct tissue reactions

In higher animals, sensitivity to electromagnetic radiation is concentrated in very specialized organs which have become both selective and extremely differentiated. Under favourable conditions a few photons are sufficient to trigger off the visual process. It should not, however, be forgotten that in addition to these specialized organs, radiation acts in other ways on the organism. Besides their visual role, the eyes have another very important function. Light is not only used to explore the outside world and to respond to it by appropriate motor reactions, it also has a special function in controlling, by means of the internal secretions, the humoral equilibrium of the animal; and in particular its sexual rhythm.

The epidermis absorbs a large part of the photons which reach it and this is appreciated as the agreeable sensation of being warmed by the sun. It is mainly the infrared, to which the superficial layers of the skin are relatively transparent, which can reach heat-sensitive nerve endings. The energy threshold necessary to initiate this sensation of heat is of an order of size incomparably greater than for vision (the ratio is of the order of 10^{11}).

Another equally well-known phenomenon is sunstroke, which can effect both the light skin of white and the pigmented skin of black races. It has been known for a long time that this is due to the ultraviolet. This acts on certain cells deep in the epidermis which release chemical substances and these diffuse through the dermis and dilate the blood vessels, producing the characteristic redness of the skin (erythema). The action spectrum of erythema (Fig. 46) shows a very pointed maximum which in classical terminology is at $\lambda_m = 296 \cdot 7$ nm, and in photons at $296 \cdot 9$ nm. This is the limit of UV which filters through ozone, which explains the considerable variations of erythema producing efficiency of the sun according to seasonal and meteorological conditions. If the erythema is mild, the red colour disappears

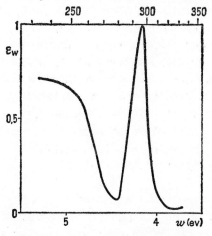

Fig. 46 Action spectrum of erythema

in several days, and the skin is left brown. There is also an immediate browning without erythema, caused by the conversion of melanin precursors in the skin under the action of UV of longer wavelength, with an action maximum of about 340 to 350 nm. Probably the action spectrum of erythema does not correspond to absorption by any definite constituent of the epidermal cells, but to the fact that the dead and cornified cells which constitute the outer layers of the skin stop all UV of shorter wavelength than 297 nm. This protects the living cells beneath from the harmful effect of short UV radiation.

As for most of the effects of UV, photosensitivity to erythema can be produced by various substances, and also photorestoration by intense visible light. Although the medical aspects of

147

the problem are outside the scope of the subject, it is worth mentioning that *carcinogenesis* that is, the production of skin cancer by a very strong dose of UV of wavelength less than 310 nm, is also susceptible to partial photorestoration by visible light.

There is a direct stimulation by light of the sphincter of the iris in certain amphibia (frog, salamander) and fish (eel, perch, salmon). The nervous system is not involved, since the phenomenon persists if the iris is excised and removed from the eye. In the eel the action spectrum has a sharply pointed maximum at about 500 nm, and a second peak in the violet. Possibly this is due to rhodopsin. In the frog the threshold of contraction of the pupil is only 10 times the absolute threshold of the cones, so it seems that this might well be a biological protection mechanism. The rods are protected by the migration of epithelial pigment. In mammals, the pupillary reflex is controlled by the retina and in man especially by the foveal cones.

Another direct action of light consists of change in skin colour. The *chromatophores* of cephalopods, for example, consist of a sac containing a dark liquid pigment and surrounded by smooth muscle. When these muscles contract, the sac is transformed into a disc whose diameter becomes 60 times that of the resting sac. The colouring agents are melanins, carotenoids and guanine. The action of light is sometimes mediated by the nervous system, but there is also a direct effect; in the sea urchin, the action spectrum in quanta has a maximum at 468 nm and it seems that photosensitivity resides in the chromatophore itself. In the frog there is a hormonal action, persisting even when the eyes of the animal are removed. It is not known where the photoreceptors acting under these conditions are situated. In the eel there seems to be both a nervous mechanism and a direct action. Knowledge of these phenomena is very rudimentary.

In many invertebrates, neurons, which do not have a true visual function, show changes when they are illuminated. The

148

giant visceral ganglion cells of the mollusc *Aplysia* for example are not photoreceptors because they are normally protected from direct light by their position in the body of the animal, but they present in spite of this the attributes of photoactivation.

An interesting case is that of direct ultraviolet irradiation of nerve cells. The resting potential and the action potential decrease (the spikes decrease in amplitude and the conduction rate of the nerve impulse drops). The action spectrum starts at about 310 nm and passes through a maximum of about 280 nm. The axon is the most sensitive part and it is possible that UV acts by modifying the conductance of Na^+ ions, and cell membrane permeability. In the frog and the crab, the loss of excitability from nerve fibres due to irradiation with UV is partially restored by photoreactivation in visible light. The phenomena of photosensitization are also seen, the nervous tissue impregnated with dye becoming excitable under radiation which is absorbed by the dye.

Many other observations could be described, evidencing a direct action of radiation on non-visual organs. The sensitivity of these reactions is usually low compared to that of the retina. From time to time press reports appear of quasi-visual functions attributed to the skin of the forehead or the fingers, of specially gifted subjects who are able to recognize colours or even writing. These examples of 'paroptic' vision would seem to have more in common with conjuring or spiritualism than science, and although occasional informed persons have taken some interest in them it would be wise to regard these reports with suspicion.

8.2. Photoperiodicity in animals

It has always been known that most animal species live according to a strongly marked seasonal rhythm. For example, in birds in the temperate region of the northern hemisphere, reproduction takes place only in the spring or summer, when temperature and abundance of food are favourable to the birth of young. Similarly moulting and migration occur at fixed periods.

In plants, Garner and Allard (1920) showed that flowering is linked to day length rather than to temperature, and suggested that periodic phenomena in animals might be due to the same cause, which they called *photoperiodicity*. From 1925, Rowan examined this problem experimentally on Canadian migratory birds. Working on the assumption that migration was triggered by hormonal activity of the reproductive organs, he showed that in spite of very low temperature gonadal activity could be induced in caged birds by artificial illumination. In France Benoit showed from 1937 onward that the photosexual reflex in the duck has a hypophyseal relay, since it is abolished by hypophysectomy, and that on the other hand it can be triggered by direct photostimulation of the hypothalamus. In California, Wolfson, beginning in 1942, completed the study of this aspect of bird migration and showed that migration was preceded by various physiological modifications, other than sexual activity of the gonads, namely, increased activity of the pituitary gland, formation of fat under the skin and in the peritoneum, and an increase in total weight.

This problem is much more complex than appeared to early experimenters. Since birds exhibit photoperiodicity most clearly, the main results which have been obtained will relate to them and in particular to birds of the northern temperate zone.

In Rowan's theory, change in day length is essentially what induces hormonal and metabolic changes in the bird. An objection immediately becomes apparent: many migratory birds spend the winter in equatorial regions where day length is almost constant, or in the southern hemisphere where day length decreases after December 21st. In this case an internal rhythm must exist which would govern the time of migration and reproduction, light serving only to maintain this natural clock in phase with the environment.

To prove this theory, six groups of the same bird species were subjected to different conditions of illumination. The first was placed in natural illumination and the others in artificial illumination of 9, 12, 15·5, 20 and 24 hours duration daily, lasting for a year, beginning in December. In the last three groups, compared with the natural control group, an early appearance and disappearance of the physiological signs enumerated above was found. For the 12-hour group, these signs appeared slightly earlier and were prolonged considerably. In the 9-hour group, the appearance of these signs was delayed by two months, and they persisted to the end of the year.

Similar experiments started off at a different time of the year have showed that the diurnal photoperiod acts on the regulation of the complete cycle and not only as a trigger for reproductive activity. In particular there is a refractory state which starts in July to August and ends, depending on the species, between October and December, during which the long days have no physiological action. This explains the behaviour of migrants which spend the winter in the southern hemisphere. The appearance of short days is necessary for the refractory state to end and only then will reproductive activity recommence. An artificial succession of short days and long days can stimulate two or three complete cycles per year in some birds. The natural clock is then not only thrown out of phase but out of control: it goes two or three times too fast.

An interesting question is whether or not there is summation of the hours of light and dark in the day. This is not true for the *preparatory* phase, which normally occurs at the end of the summer and in the autumn, and which is controlled by the presence of darkness (D) for at least 12 consecutive hours. Thus the sequence 4L-8D-4L-8D has no effect, although 8L-16D has; 6L-6D-6L-6D has no effect and 12L-12D is just effective. Excessive day length would, however, be inhibitory and a cycle such as 16L-16D would be ineffective although the night is longer than the 12 hours of threshold. On the other hand in the *progressive* phase which follows the preparatory phase, it is the presence of light which constitutes the active stimulus and short periods of illumination are added together in the 24-hour period. In this case continuous and excessive duration of darkness is inhibitory. The mechanisms are therefore quite different in the two cases.

In the male bird it is chiefly testicular growth which is governed by light and in the female, only the initial stage of ovarian development shows a photoperiodic response. In many species (notably the sparrow) ovulation requires other stimuli (presence of the male, materials and site for the nest).

In the domestic fowl, selection has considerably modified photoperiodic behaviour, which was originally of a tropical type. Here day-length acts differently from the species of temperate regions. Egg production is not limited to a given period, but varies with the season and, as is well known, winter production can be increased by artificial illumination which extends day-length. Sexual maturation of young pullets is equally hastened by supplementary light.

Leaving the subject of birds, we will now quickly review photoperiodicity in general. Some cases of light-controlled rhythms have been reported in invertebrates other than insects, one of the most curious being that of the worm *Platynereis dumerlii* where the light of the full moon seems to be a factor in synchronization. In insects photoperiodic mechanisms are varied and interact with environmental factors (temperature, food, population, density) to control especially metamorphosis, as well as sexual form. In the silkworm, light acts directly on the brain through a small transparent aperture. In some aphids parthenogenesis characteristically appears in spring and summer populations and oviparous females and males in autumn and winter. It is difficult to see how light can control the appearance of males other than by modifying a tendency of the eggs to lose an X chromosome during mitosis.

According to Lees (1966), in the aphid *Megoura viciae*, the action spectrum of the phenomenon has a maximum of about 460 nm, when light interruptions of a dark period occur towards the beginning of the period. It is displaced towards the longer wavelengths for interruptions later in the period. Probably changes occur in the active pigment (which is perhaps not present in the compound eye, but directly in the brain).

There is little information about photoperiodicity in amphibians and reptiles, and hardly more about fish, in spite of various experiments on migratory species. It seems, however, if the temperature of the medium is high enough, that light periodicity may be an important sexual factor in some fish. The bitterling and the stickleback can be classed among the 'long day' species and the trout among the 'short day' species. In mammals, on the other hand, numerous studies have shown similar phenomena to those described in birds. The reproductive cycle in females is induced or accelerated by long days in some species (horse, hare, ferret) and by short days in others (goat, sheep). Domestication has interfered with photoperiodicity in bovines and pigs. In the ferret, the 'long day' apparently signifies chiefly the absence of continuous darkness of more than 12 hours, so that the 2L-10D-2L-10D cycle acts as a long day although the light has only lasted for 4 hours. This does not occur in birds.

The physiological mechanism of photoperiodicity has been specially studied in the duck by F. Benoit and his collaborators. The essential photoreceptor is the retina, from which three types of nerve fibre pass in the optic nerve. The first, which is specifically visual, terminates in the cortex after relay by the lateral geniculate nucleus; the second initiates the optical reflexes (diameter of the pupil) and terminates in the pretectal area; finally, the third constitutes the *photo-neuro-endocrine* circuit which reaches the *hypothalamus* (Fig. 47). The hypothalamus secretes a product which is diffused in the circulation to the adenohypophysis (lobes of the pituitary). This secretes in its turn gonadotropic hormones which act on development of the testicle. In fact, this schema of retino-neuro-endocrine circuits is still extremely hypothetical and now the tendency is unfavourable to the direct pathways between retina and hypothalamus shown in Fig. 47.

It is not known for certain whether the retinal visual cells (cones and rods) also constitute photoreceptors for the two non-visual types of fibre. The action spectrum of the pupillary

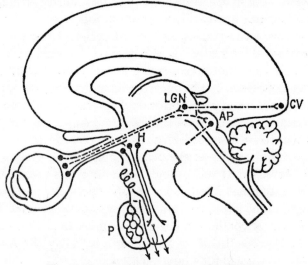

Fig. 47 Photo-endocrine circuits, after Sharrer
—·—·—·— visual circuit (LGN, lateral geniculate nucleus;
CV, visual cortex);
——————— optical reflexes circuit terminating in the pretectal area, AP;
———— neuro-endocrine circuit; H, hypothalamus; P, pituitary

reflex seems to differ a little from the photopic luminous efficiency V_λ in the sense that the short wavelengths of the visible and UV act mainly on the pupil. For the photo-endocrine effect, the action spectrum is, on the other hand, displaced towards the longer wavelengths (maximum about 650 nm). It is possible that the transparency of the eyelids plays some role and that the birds may protect themselves by shutting their eyes during part of the experiment. It also seems possible that light may act directly on the hypothalamus and here again the

154

transparency of the tissues of the head could explain the maximum action of orange and red radiation. Light intensity does not obey completely the law of reciprocity, and the optimum value is relatively low (from a few lux to about a hundred, depending on the species studied).

In mammals, the mechanism appears to be analogous to that seen in the duck, but the retinal photoreceptors predominate. Section of the optic nerve totally suppresses the photoperiodic response, and it is impossible to substitute for it by direct illumination of the hypothalamus. In the rat, fibres from the optic nerve terminate in the anterior part of the hypothalamus, just behind the optic chiasma. In mammals the pineal gland may also play a part in photoperiodicity mechanisms, but this is as yet not proven. In cybernetic terminology, the *feed-back* which clearly operates in the mechanism of pupillary contraction (which I have purposely not mentioned since there are several recent good reviews of this subject) must also play a part in photoperiodicity, and multiple controls of this type probably occur. In particular the thyroid and adrenals show definite annual cycles.

Although the annual rhythm of animals, and in particular their sexuality, is clearly governed by light, the absence of it does not necessarily lead to sterility. Benoit kept ducks in total darkness for 11 years; they nevertheless reached sexual maturity, although late, and subsequently showed short and irregular cycles of testicular activity. In man, slight metabolic changes have been observed in blind people (elimination of water, hypoglycaemia, etc.).

Permanent exposure to a strong light can on the other hand induce degeneration in the visual system of animals.

8.3. Circadian rhythms

In addition to the annual seasonal rhythm, most living things shows an approximate 24-hour periodicity. This rhythm was called *Circadian* (Halberg, 1959) and has stimulated a considerable amount of experimental work. It exists in various degrees in all animals and plants. The initial fundamental question is whether this rhythm is acquired and due to environment and habits, or whether it is innate with daylight in particular having only a synchronizing role.

The existence of a hereditary factor seems certain. It has been possible to maintain the fly *Drosophila* in constant physical conditions unchanged over 15 successive generations without eliminating the diurnal ryhthm, which must be linked to a 'biological clock'. Obviously, in the absence of an external fixed period, this clock is easily put out of control, but it can also be reset accurately to the hour, a single period of 10 hours light being sufficient to re-establish synchronism. This suggests that the biological clock is not of sinusoidal type, but rather of the oscillation of relaxation type, or better still operates according to certain non-linear mechanisms the mathematics of which are complex.

The detailed study of rhythms is a long and difficult statistical operation, based on Fourier's harmonic analysis. By this means a *frequency spectrum* (periodogram) can be established and the significance which can be attached to the natural frequencies so observed can be evaluated. These elaborate methods, perfected in astronomy and meteorology, are little by little penetrating biological fields since only by these means can objective conclusions be reached.

The crab *Uca pugnax*, for example, in its natural habitat shows colour changes over a 24-hour period, due to melanophores which make the legs darker by day than by night. A tidal rhythm (24·8 hours) is also found but to a lesser degree. If the crabs are placed in an aquarium after their eyes have been removed, harmonic analysis of the rhythm of the melanophores during 39 days gives a periodogram whose maximum is about 23 hours, but with marked variations from one subject to another.

156

The mouse has been investigated extensively and its Circadian rhythm can be objectively evaluated by measuring its urine corticosteroid concentration. When illumination is changed from 6 a.m.–6 p.m. to 6 p.m.–6 a.m. it takes about 14 days before the diurnal rhythm is reversed. The mouse shows a diurnal rhythm for many phenomena: rectal temperature, blood eosinophils, serum corticosteroids, liver glycogen, etc. For the first month after bilateral enucleation of the eyes, the periodism always deviates from the 24 hours (23·3 hours) value, but after several months there is a partial return to synchrony under the influence of periodic phenomena which are usually secondary with respect to the eyes.

Birds have been the subject of numerous studies. If they are submitted to continous illumination, the biological clock begins by advancing. In the starling, the rhythm remains at 24 hours for 30 lux, whereas it is shortened to 23 hours at 300 lux. A prolongation of intense illumination progressively decreases the Circadian rhythm and the clock finally stops altogether. If at this moment the animal is placed in the dark or in a weak light, the 24-hour rhythm reappears spontaneously. This at least, is what happens with diurnal animals; in nocturnal animals the biological clock slows down if they are illuminated excessively.

Attempts have been made using birds to study the establishment of the natural rhythm by allowing the animal itself to control it. Canaries are placed in a cage and a contact which puts out the light is closed when they alight on the perch to sleep. When they leave the perch the light comes on again. On 47 birds, 35 established a Circadian rhythm in less than a month. The periodicity differed with each bird (ranging between 23 and 24·5 hours). The time of waking was found to be more regular than that of going to sleep. Certain drugs shortened the period.

In the monkey the establishment of a Circadian rhythm has also been followed. Young separated very early from their mother are reared for three years in the dark, apart from one hour of diffuse light daily to prevent degeneration of the retina.

157

The rhythm is evaluated by general and spontaneous activity. Feeding time has no effect, but the time of the hour of illumination has a strong effect; a sub-period of 12 hours can be induced by 2 half-hours of illumination separated by a 12-hour interval.

In man, the establishment of natural rhythms has been studied in nursing infants. The innate periodicity of the sleep-waking rhythm appears very quickly, starting in the first week (with an inversion of phase due doubtless to hunger). Apparently

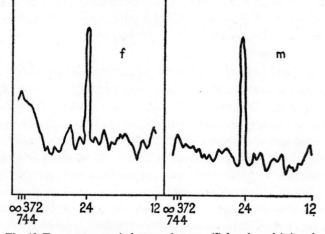

f   m

∞372   24   12 ∞372   24   12
744   744

Fig. 48 Temperature periodograms in man: (f) female and (m) male
(the hours are measured from right to left)

it is not 24 hours exactly, but more like 24·4. Then (in the third week) a diurnal variation appears in the electrical resistance of the skin, showing a morning maximum towards 10 a.m. At about the 20th week several rhythms are established: temperature, heart rate, urine excretion (diurnal maxima and nocturnal minima); at 6 months a renal rhythm is observed acting on sugar and Cl^- ions, the infant reaches 18 months before urinary phosphates and creatine also show a Circadian periodicity. All these periodicities are established more slowly in the premature than in the normal infant.

By way of example, Figure 48 shows the Circadian rhythm of

158

rectal temperature in the adult. The periodogram shows besides the definite peak at 24 hours, a secondary peak related in women to menstruation but which also seems to be present in men, although it is much less marked. The rhythms of leucocyte count, 17-ketosteroid excretion, hypophyseal adrenocortico-tropin concentration, blood and urine 17-hydroxy-corticosteroid concentrations, and lastly intra-ocular pressure, have also been measured in man. For example the maximum plasma corti-costeroid concentration occurs during sleep (sleep verified by electroencephalogram) and precedes the beginning of diurnal activity. Mrs. Radnot has tried to investigate the effect of light on the blood eosinophil cycle. At the end of 20 minutes illumina-tion, the concentration starts to fall and the minimum is reached after 2 hours. The action spectrum seems almost identical to the photopic efficiency V_λ, but the measurements are less accurate. In man, in addition to the natural cycle due to light, there are very important effects due to environment and routine, and pyschological factors react on the cortical system. Light maintains the subject in a state of alertness and this can easily be seen by looking at the pupillary contraction reflex. Fatigue and approaching sleep diminish the pupillary response to light variation, and when the subject is going to sleep, the pupil is constricted and barely responds to light. If the onset of sleep is interrupted, by a loud noise for instance, the pupils dilate and recover their normal reflexes to light. For man, the day is not only the time when physical stimulation by light is at a maximum, but also the time when this stimulus can act most on the senses, this being probably due to the activity of the sympathetic innervation of the hypothalamus. Attempts to replace the 24-hour sleep-wakefulness rhythm in man by cycles of altered length have had some success for cycles of slightly altered length, up to 21 or 27 hours, but none for cycles such as 12 or 48 hours. For instance Meddis (1968) found no sign of adaptation to a 48-hour routine for seven subjects studied for a period of two months: the normal diurnal body temperature

159

rhythm was unaffected, as were all psychological and physiological tests. Records of electroencephalogram and eye movements made on one of the subjects during sleep show a striking difference: during the experimental night, the rapid eye movements periods (indicators of dreaming) are clustered in the beginning of sleep, while in the control night most periods of dreaming are near the end of the sleep, as occurs normally. This 'dream deprivation' is perhaps an explanation of the failure of adaptation to the 48-hour day.

Probably in the higher animals and in man, the interpretation of Circadian rhythms and their relation to light is not yet close to a satisfactory solution. It is reasonable to direct, at least to begin with, experimental effort to the study of lower animals. Some results in this direction have already been obtained. Thus when an organism with a Circadian rhythm is exposed to a cycle of illumination of 24 hours but differently phased, a progressive dephasing of the old rhythm towards the new occurs after a disordered period. This is well known to aeroplane travellers suffering from a change in longitude. Pittendrigh and Bruce (1957) discovered that in *Paramecia* a similar dephasing occurs under the action of one or more flashes of intense light. These have little effect if they occur during the day but they induce a phase retardation or advance depending on whether they are applied during the first or the second half of the night. This phenomenon is of interest since it is better suited to the study of the variation of the physical parameters of the light flash than the complex effects in the higher animals. The action spectrum appears to be identical to the total absorption spectrum of the cell. The chemical intermediates are still unknown. However, in the case of photosynthesis in *Acetabularia major*, it has been shown that actinomycin D inhibits the biological clock; it is known that actinomycin D is a specific inhibitor of the synthesis of ribonucleic acid from DNA, which suggests that ribonucleic acid plays a part in the mechanism of the biological clock, at least for the organism in question.

REFERENCES

ASCHOFF, J., ed., *Circadian Clocks*, Amsterdam, North-Holland Publ. Co. (1965)

BENOIT, J., Gen. and comp. endocrinol. Suppl. 1, 254 (1962)

DANILEVSKII, A. S., *Photoperiodism and seasonal development in insects*, Edinburgh, Oliver and Boyd (1965)

GARNER, W. W. and ALLARD, H. A., J. Agr. Res. 18, 553 (1920)

HALBERG, F., HALBERG, E., BARNUM, C. P. and BITTNER, J. J., *Publ. 55 Am. Soc. Adv. Sci.* Washington, 803 (1959)

LEES, A. D., J. Insect Physiol. 10, 92 (1966)

MEDDIS, R., Nature 218, 964 (1968)

PITTENDRIGH, C. S. and BRUCE, V. G., in *Rhythmic and Synthetic Processes in Growth* (D. Rudnick, ed.), Princeton University Press (1957)

RADNOT, M. in Photo-neuro-endorine effects in circadian systems, Ann. N.Y. Ac. of Sci, 117 (1964) (WHIPPLE, H. E., ed.)

ROWAN, W., Nature 115, 494 (1925)

WOLFSON, A., Condor 44, 237 (1942)

CHAPTER IX

Bioluminescence

9.1. Luminous organs in animals

To round off this very condensed review of the effects of light in the animal kingdom, the inverse phenomenon, that of light production by organisms, will be briefly considered. It occurs randomly in many invertebrates and in fish, but no example is known in terrestrial vertebrates. So called 'phosphorescence' of cats' eyes is simply a reflection of light from the back of the eyeball, well known to motorists who pass a cat or a rabbit in the darkness. This reflection is specially noticeable in animals who possess a *tapetum*, i.e. a region of the choroid which reflects light instead of absorbing it. This mirror is built with crystalline layers and it is thought to aid sensitivity by increasing the intensity of light in the photoreceptors. Such a tapetum exists also in many invertebrates and is the origin of the glow exhibited by some insect eyes. The reflection comes from a stratified layer just underlying the rhabdoms and functioning optically as an interference filter. It is interesting also to note that the compound eyes of many insects are coated externally with a regular array of *corneal nipples* which act as an anti-reflection coating and reduce the corneal reflection that might otherwise attract predators.

'Stars' seen after a blow on the eye is also sometimes wrongly attributed to light emitted by this organ, and in one lawsuit it was reported that the plaintiff pretended to recognize his assailant in the dark by light given out following a blow on the eye. These '*phosphenes*' have, of course, no luminous basis. In the same way, *Purkinje's blue arcs*, which are seen when watching

162

a small red light in the dark, and which follow the retinal path of the fibres between the fovea and the optic nerve, were once thought to be due to retinal luminescence following the visual stimulus. This phenomenon is also a phosphene, but in this case an electric one, due to induction by the fibres of the visual cells which they pass near. The only biological light produced

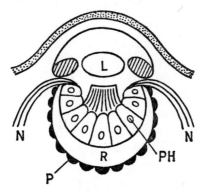

Fig. 49 Photophore of shrimp, after Chun
L, lens; N, nerve; P, pigment; PH, photophores; R, reflector

by air-breathing vertebrates is the fluorescence emitted by UV-irradiated organs, especially the lens of the eye.

The anatomical structure of the luminous organs of certain animals bears a curious resemblance to the eyes of other species. Some photophores in shrimps (Fig. 49) recall the dioptric eye of cephalopods and vertebrates, although the shrimp itself has a quite different compound eye. Conversely, the decapod *Acanthephyra* (Fig. 50) has a luminous organ which recalls a faceted eye, except that it has no facets on the cornea and the pigment layer is reflecting. In the fish *Bathylychnops exilis*, in addition to the normal dorsal eyes, there are two small ventral swellings which were at first taken for photophores but which are in reality secondary eyes with lens and a retina with four layers of rods, like the dorsal eyes.

The parallelism of eyes and photophores has, of course, given rise to pseudo-philosophical digressions on the evolutionary significance of bioluminescence. The one case where these considerations are of interest are the crustaceans and deep-ocean fish. Usually the eyes of crustaceans who live in the darkness at great depths degenerate if they have no photophores,

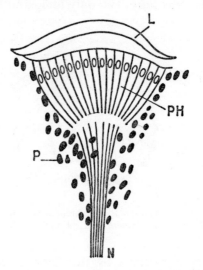

Fig. 50 Decapod photophore, after Kemp (same labelling as Fig. 49)

whereas, on the other hand, the majority of luminescent copepods are blind.

9.2. Mechanisms of bioluminescence

The mechanism of light emission by animals was elucidated by R. Dubois in 1885. An enzyme secreted by the animal (*luciferase*) catalyses the oxidation by air of a substance (*luciferin*). This oxidation process liberates a large amount of energy, an intermediate product becomes excited and then returns to its fundamental state by emitting fluorescent light.

The glow-worm and firefly are known by everyone, but there are many luminous animals distributed at random through the classification. Often one of two closely related species emits light and the other does not. Classically, luminescence is interpreted as a sexual signal. When only one sex is luminous it is the female and when both are, a different code of light sequences makes recognition possible. In tropical regions, hundreds of animals can sometimes be seen on the same tree, synchronizing their emission of light.

The general problem of bioluminescence cannot be dealt with in detail in this volume. Essential phenomena will be illustrated by a few examples. There have been few good spectral measurements of the energy density as a function of the wavelength (the only mode of representation which most naturalists seem to know). It is usually impossible to construct the emission curve from it with the conventions we have been following, that is the number of photons emitted in equal intervals Δw (in elementary photon energy). Unless otherwise specified the values of λ_m given are thus relative to the energy emitted in bands of equal width $\Delta\lambda$. Unless the spectrum is very narrow, the maximum number of photons with division into Δw is different and displaced towards the red. No value must therefore be given to coincidence between λ_m and the absorption maximum of a visual pigment (which itself has a real significance) to deduce that the animal emits light which it can see. The problem of biological adaptation is more complicated, and we shall return to it later.

We will start at the bottom of the scale with luminescent bacteria. Their emission is intracellular and usually continuous. The λ_m value depends on the species, for example it is 465 nm for *Photobacterium phosphoreum* and 490 nm for *Achromobacter fischeri*. The temperature effect is also different, luminescence being maximal at 20° C and 28° C for the two species respectively. Luminescence is abolished in both at about 40° C, and at about 5° C in the second species while the first still emits light at this

165

temperature, although it is 5 times less than the maximal value. Bacterial luciferin and luciferase have not been isolated, but possibly the first may be a reduced flavine mononucleotide ($FMNH_2$) which in the presence of a long-chain aldehyde is oxidised to FMN. As for the luciferase of *Achromobacter*, its absorption spectrum shows a maximum of 370 nm, besides the classic protein band at 280 nm. It is difficult for the determinists to explain the biological significance of bacterial luminescence, except perhaps in cases of symbiosis with fish. Among the protozoans, some marine flagellates are strongly luminescent.

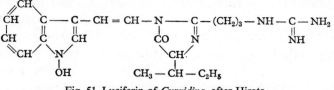

Fig. 51 Luciferin of *Cypridina*, after Hirata

Gonyaulax polyedra exhibits a diurnal rhythm of luminescence, which manifests itself as brief light flashes of about 0·1 second duration in the interior of small rhombohedric crystals of 0·3 to 0·6 μ in length. The presence of Na^+ or K^+ ions is essential, as well as oxygen, for this luminous emission.

In the medusa *Aequora aequora*, luminescence is apparently governed by a very special mechanism not requiring oxygen, but requiring water and Ca^{++} ions. There appears to be no enzyme and the light is emitted by a protein with a molecular weight of about 35,000.

The creature in which luminescence is best known is a Japanese crustacean, *Cypridina hilgendorfii*. The luciferin has been extracted, purified, and its chemical structure elucidated (Fig. 51). The luciferase is a protein with an absorption maximum of 277 nm and a molecular mass of about 50,000. These compounds can be made to react *in vitro*. There is a simple first-order reaction, influenced by pH and temperature. The presence of Cl^- ions increases the intensity of the light emitted and this has been carefully studied spectrally, and the usual kind of curve can be constructed (Fig. 52). Whereas the λ_m with the classical conventions is about 460 nm, in the curve illustrated here the maximum is 470 nm. The quantum efficiency, defined by the ratio of the number of photons emitted to that of the number of luciferin molecules oxidized, is about 0·3. Urea and cyanides inhibit luminescence, even at low concentrations.

Among luminous fish, *Apagon ellioti* is of interest, since the luciferin and luciferase, extracted from a luminous organ, are

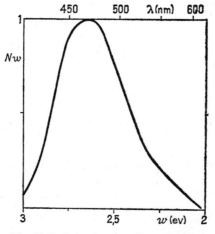

Fig. 52 Emission by *Cypridina*, after Sie

identical to those of *Cypridina*, in spite of the separation of these animals on the evolutionary scale. The four emission spectra resulting from mixing the two luciferins and the two luciferases have been recorded, and are identical. If we accept that vision in *Apagon* is mediated by a pigment obeying Dartnall's nomogram, we can, by integration methods similar to those which we have used to define adaptation of a pigment to light penetrating the sea, find which λ_m is best adapted to luminescence in Apagon. This is found to be 473 nm which shows clearly that the classic analogy is faulty (460 nm).

167

It has also been possible to extract, purify and identify the intermediate products of luminescence in the firefly *Photinus pyralis* and these are probably the same for all *Coleoptera*. Luciferin has been synthesized as two possible stereoisomers (Fig. 53), in which the COOH radical is on one or other side of the plane of the figure. Only the right-handed isomer is active in light production. If E denotes luciferase, that is, the enzyme necessary for light emission, and L dehydroluciferin (Fig. 53), luciferin, LH_2, combines with adenosine triphosphate in the

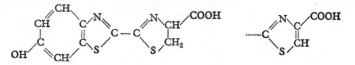

Fig. 53 Luciferin (left) and dehydroluciferin (right), after White

presence of Mg^{++} ions, and eliminates inorganic pyrophosphate PP:

$$LH_2 + ATP + E \rightleftarrows E\text{-}LH_2\text{-}AMP + PP$$

This is followed by oxidation of the adrenylate of luciferin:

$$E\text{-}LH_2\text{-}AMP + O_2 \rightarrow (E\text{-}LO\text{-}AMP)^* + H_2O$$

and the excited molecule (indicated by the asterisk) returns to its fundamental state emitting a fluorescent light: this oxidation reaction is also catalysed by luciferase E which therefore plays a dual role.

The emission spectrum of *Photinus pyralis* has been measured *in vitro* at pH 7·6 and the curve, transcribed according to our convention (Fig. 54), has a maximum at 565 nm (the classic λ_m is 562, the difference is small here because the curve is peaked). The quantum efficiency is nearly unity, that is one photon is emitted per molecule of luciferin oxidized. If the medium becomes acid, the curve is displaced towards the long wavelengths and the quantum efficiency drops; for pH 6·0, the standard λ_m would be 614 nm. Temperature increase has an analogous effect.

Probably all fireflies function by the same process, but small alterations in the amino acid of the enzyme modify the colour, the classic λ_m of which varies between 550 and 580 nm.

Light is emitted by fireflies in flashes which follow one another at a frequency which varies from one every two seconds to 20 per second, the rhythm being characteristic of the species. It is

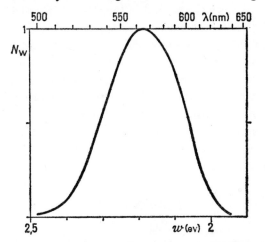

Fig. 54 Emission by *Photinus pyralis*, after Seliger and MacElroy

thought that the nerve impulse which controls light emission acts by liberating pyrophosphate into the cytoplasm. This in its turn causes the liberation of the enzyme (previously inhibited from acting), which acts on luciferin: this mechanism is however still hypothetical. According to J. and E. Buck (1966), the hypothesis of visual feed-back is the only possible explanation of the astonishing synchrony of emission in certain tropical fireflies on the same tree.

Bioluminescence seems to be so randomly distributed in plants and animals that it has been suggested that it might be the remnant of a phenomenon ancient from an evolutionary

BIOLUMINESCENCE

point of view. It might for example have been a much more
efficient method than the present ones (by haemoglobin-type
pigments) of fixing oxygen, which in its free form was very
scarce in early times. This might have been achieved by utilizing
complex organic molecules as reducing agents, which passed
into an excited state, and thus became fluorescent; light would
have been a side-product and not the goal of the operation. When
oxygen was a rare commodity, at the start of evolution, strong
reducing systems were needed in the struggle for existence; this
meant states of high-energy excitation and the emission of
photons of short wavelength. Bacteria actually emit blue light.
The reducing power of the environment then decreased and the
energy of excitation with it; luminescent multicellular organisms
generally radiate in the yellow-green range. The philosopher
who sees in his garden the common glowworm emitting its
light is tempted to imagine that he is witnessing the last sparks
of a fossil firework, rather than the somewhat crude efforts of a
female to attract a mate.

REFERENCES

BUCK, J. in MCELROY, W. D. and GLASS, B., *Light and Life*, Baltimore, Hopkins (1961)
CHUN, C., Wiss. Ergeb. Valvidia Exped., **18**, 1 (1910)
HIRATA, Y., Tetrahedron Letters, **5**, 4 (1959)
KEMP, S., Proc. Zool. Soc. London, 639 (1910)
SELIGER, H. H., and MCELROY, W. D., Arch. Biochem. Biophys. **88**, 136 (1960)
SIE, E. H. C., Arch. Biochem. Biophys. **93**, 286 (1961)
WHITE, E. H. in MCELROY, W. D. and GLASS, B., *Light and Life*, Baltimore, Hopkins, 1961

Conclusion

Having finally arrived at the end of this rapid review of a large scientific field both reader and author share a feeling of admiration for the patience and ingenuity of research workers in this field and a hopeful attitude towards future developments. Our knowledge is very small compared to our ignorance and the possibilities are there for the young who wish to participate in this fascinating adventure. A great number of projects are being made now in various laboratories; they need to be co-ordinated and stimulated, but unfortunately at the present time the study of light is eclipsed by preoccupation with the atom or with space.

Animal photobiology and its practical applications could, however, be extremely useful. I shall give three examples only. If research on phototaxis were to lead—which is not impossible—to the separation of male and female spermatozoa, the consequences in terms of human sociology and animal rearing would be far-reaching. Annual and Circadian rhythms condition all our activity, and pharmacological advantage might be taken of these. For example the resistance of mice to death from an injection of ethyl alcohol is maximal about two hours after the start of the diurnal illumination cycle, and minimal at the end of the day or the beginning of night. In man it also seems that drugs act differently depending on their time of ingestion. Individual variations occur as well and these seem to be of genetic origin, as is witnessed by the identical behaviour of identical twins. Study of mono- and dizygotic twins has shown that the curve of the Circadian rhythm of the pulse seems to be significantly genetically influenced. In our civilization, where artificial light replaces or supplements sunlight, a knowledge of

171

the rhythms linked to illumination is essential; in this field we are totally ignorant at the present time. Finally, the possibility of studying the cones in the human retina *in situ* will lead in the future to the better understanding of the anomalies of colour vision and perhaps their correction. It is probably not widely appreciated that about 8 per cent of males are abnormal in this respect, and in modern life where colour and optical signals are so important, such a deficiency is a hindrance in many types of job.

'You are in the trade, my dear Sir', said Molière, and although this is obviously true I sincerely believe that in the future an increased knowledge of photobiology will be essential to the technical and social development of man. This is probably true for both animal and plant photobiology. Although on the latter subject I wish to say nothing except to observe that photosynthesis has constituted for ages past and still remains the largest industry on our planet. In the Bible, light was created before the earth and scholars do not deny this but attribute to the sun's ultraviolet rays an essential role in the genesis of life. Efforts to increase understanding of the biological function of light will certainly be rewarded in generations to come.

Index

INDEX

177